# The Six-Inch Lunar Atlas

Don Spain

# The Six-Inch Lunar Atlas

## A Pocket Field Guide

Don Spain
Louisville Astronomical Society
Louisville KY
USA

ISBN 978-0-387-87609-2 e-ISBN 978-0-387-87610-8
DOI 10.1007/978-0-387-87610-8
Springer Dordrecht Heidelberg London New York

Library of Congress Control Number: 2009928847



Printed on acid-free paper

Springer is part of Springer Science+Business Media (www.springer.com)

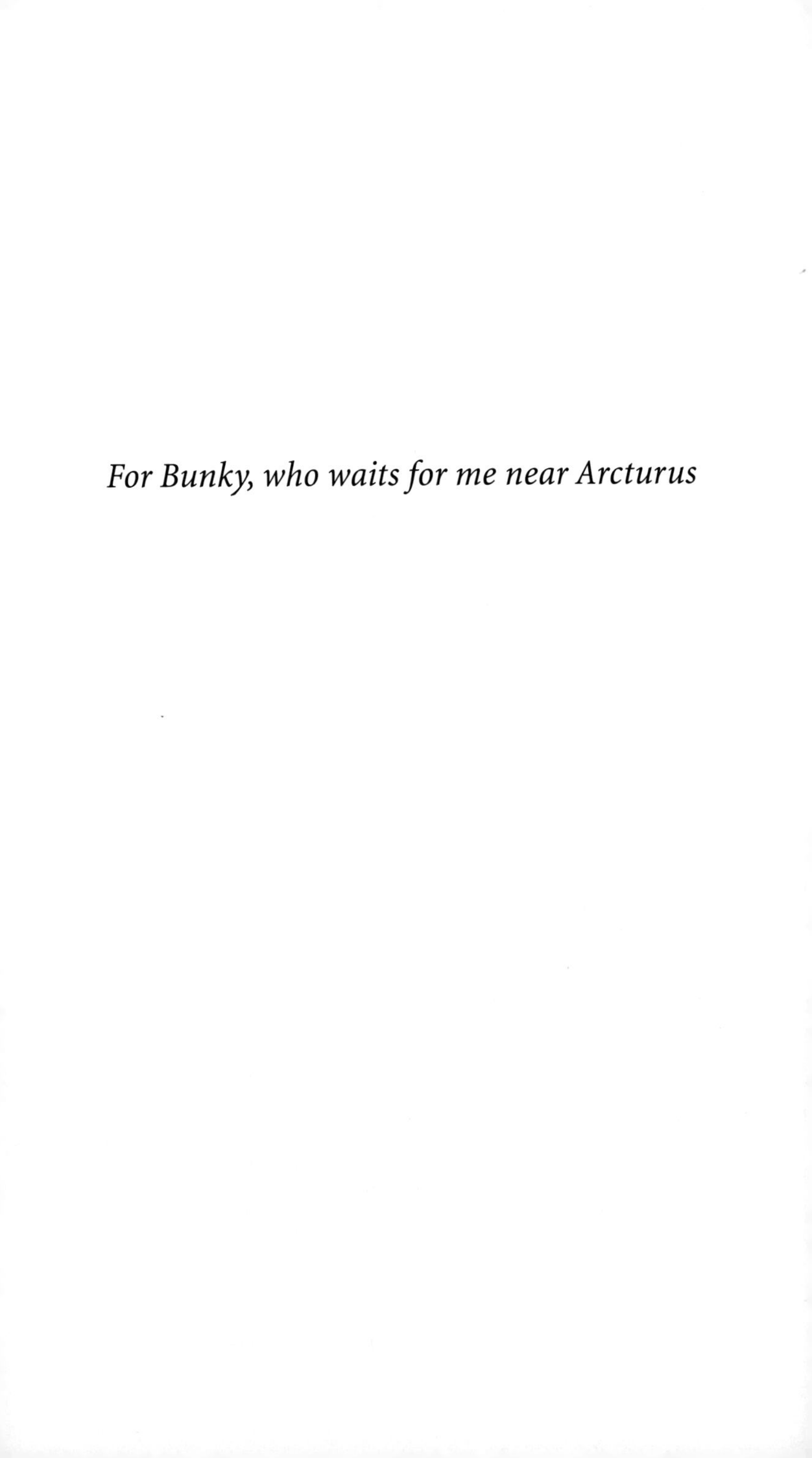

*For Bunky, who waits for me near Arcturus*

# Preface

November 26, 1922. Howard Carter breaks open a hole into the tomb of the Egyptian pharaoh Tutankhamun. His benefactor, Lord Carnarvon, asks, "Can you see anything? Carter answers, "Yes, Wonderful things." His reply is the same answer I would have given if anyone would have asked me the same question on January 1, 1958, when I first viewed the Moon through my first tiny telescope.

It was a cold winter's night, and Selene, the ancient Greek name for our Moon, was beckoning to me. Six nights earlier my parents had presented me with a small spotting scope for a Christmas present. Finally, the skies cleared, and I got the chance to use it to view the heavens. And there she was, riding high in the eastern sky. In her glorious gibbous phase that night, the Moon awaited my arrival. A brilliant white disk, almost like it was just freshly painted, she slowly floated into the field of view. I soon noticed the large, dark, smooth areas and great ray patterns and various other markings. Then along the eastern edge of the disk, craters! Wonderful craters!

That small telescope had a 40 mm objective, magnified 30 times, and was mounted on a tabletop tripod. However, it was more than enough for me to embark on my journey into the world of amateur astronomy, and in particular my lifelong love affair with Selene. I wanted to know what craters and formations I was observing. The next day I looked up astronomy and the Moon in an old set of encyclopedias we had, and I carefully cut out the simple map of the Moon. In the next 3 months I studied the Moon every clear evening. I learned the location of all the major seas and mountains and the 50 craters that were identified on that map.

Over time I progressed to larger and more powerful telescopes. The amount of detail I saw increased geometrically, and I eventually read almost every book our public library had on astronomy and their few books about the Moon. Over the years, I have gathered quite an extensive library of lunar books, maps, and atlases, including a good size collection of NASA atlases and books.

You may reasonably ask why I have compiled this little atlas when there are so many other fine lunar atlases and maps readily available. There are indeed excellent photographic atlases, on-line atlases, and other easily available atlases on CD ROM. And in addition, almost every new introductory astronomy book includes a decent lunar map, and some even have brief atlases that allow the reader to identify the more prominent lunar features.

However, most of the available atlases are too large and too inconvenient for using at the telescope. To use an online atlas or a CD ROM near the telescope you must lug a laptop or notebook computer along. The small size of this atlas should easily fit into any equipment case or bag and can be used right at the eyepiece. This book is titled *The Six-Inch Lunar Atlas* for two reasons. When closed, it is about 6 inches across, and all the photographs and images in the atlas proper were obtained with a 6-inch refracting telescope. Although designed for the beginning or casual observer of the Moon, this atlas should prove useful to anyone who wishes to locate many of the major features on the Moon or to acquire a little knowledge about these formations.

While we sometimes observe the heavens alone, most of us enjoy being with others of similar interests or teaching or just showing others what is up there. My local astronomy club, the Louisville Astronomical Society, holds dozens of public, private, and school star parties every year. These observations are usually held when there is a thick crescent, first quarter, or gibbous Moon in the evening sky. The Moon is an easy target for any size telescope, and many of the persons who attend these events have never looked through a telescope before. It is always satisfying to hear "Oh, wow!" or "That's amazing!" or "Look, mom, you can see the holes in the Moon!" And most of them will ask questions about what they see. "What crater is that?" "How big is it?" "Are those mountains?" "What is that long, snakelike line beneath that crater?" "Are those cracks in the Moon?" "How deep are they?" "What is that large, dark, smooth area?" "Where did the *Apollo 11* astronauts land?" These are just a few of the many questions I am asked at these events.

There are usually only one or two members of most astronomy clubs or societies that can answer these and other questions about our Moon. My hope is that the reader of this atlas will be able to recognize many of the prominent and interesting formations and be able to tell other observers what they are viewing through the eyepiece. To do this an atlas is almost a necessity. To locate a street, town, country, or even a continent or an ocean on Earth you need a map or an atlas. An atlas is principally a collection of maps. For the atlas part of this book, there are 60 primary photographs of the lunar surface. Imaged through my 6-inch refractor and a digital camera, they cover regions of the Moon from about the third day after a new Moon right up to nearly the full Moon.

With the introduction of a dedicated digital camera for lunar and planetary imaging I began to take literally thousands of images with the big refractor. After processing with software I was amazed by their resolution and clarity. The best ones show more detail than most of my drawings, and all formations are in the correct proportion and location with the surrounding lunar terrain. Even though 6 inches of aperture is rather small these days, the images it takes with even the inexpensive cameras is impressive. I have a fine image of the great crater Copernicus that was taken under almost the same lighting conditions as a film photo taken with the Lick Observatory's 36-inch refractor in 1936. The Lick image is a great film image, but the digital image with the 6-inch lens reveals almost the same detail, resolution, and clarity.

However, first and foremost, I am a visual observer. Exploring the terminator night after night, seeking out my old friends and finding new formations is still, in my opinion, the only way to be a lunar observer. Digital and even film images are informative and are a permanent record at any particular time. They are great to study and examine but are just a frozen snapshot of recent lunar history. When observing visually I have the feeling of orbiting the Moon and exploring its surface through the eyepiece. It's as if I am looking through the porthole of a spaceship. As I view the surface below me I find it rewarding to know the names and basic attributes of the prominent and unique formations.

So take a journey to the Moon with me. Bring this atlas with you to your telescope and observe. The size or type is not important, just a desire to learn more about our sister world. You will find that as you learn the locations and names of the many and varied formations they will become your friends. You will look forward to meeting them month after month and year after year. Many will become your favorites, and you will delight in showing them off to your colleagues and friends.

My hope is that this little atlas will inspire you about Selene, and if you become as bewitched by her as I have been you will wish to learn more. There are still many unexplained formations that are waiting for you to discover. However, they are beyond the scope of this book, so I will leave it to you to search them out for yourself. A lifetime of adventure is there for those who are willing to take up the challenge. Wonderful things await you.

Don Spain
Louisville, KY

# Acknowledgements

Thanks to John Watson for his great assistance and encouragement that led to this atlas.

Very special thanks to Maury Solomon for her understanding of the delays and her extraordinary patience with me.

To Kevin Clark, my grateful acknowledgement for letting me use your beautiful picture of the full Moon for the cover image and the several finder charts in this atlas.

And to all my astronomy and non-astronomy friends and family whose encouragement is sincerely appreciated.

# How to Use This Atlas

This atlas is designed for use with almost any type of telescope. The maps and images will be shown in three different configurations. The first will be normal, or as you would see the Moon with your unaided eye, through binoculars, a spotting telescope, or scopes with a terrestrial eyepiece or an erecting prism. In this view, lunar north will be at the top and lunar east to the right. The next image will be flipped vertical compared to the first. This is the view as seen through a Newtonian, Cassegrain, or a refracting telescope without a diagonal. In this configuration lunar north is at the bottom and lunar east to the left. The next image is a mirror image of the first, or the view as seen through a catadioptric (a Maksutov or Schmidt-Cassegrain), or a refractor with a diagonal. Lunar north will be at the top, but lunar east is to the left. By using one of these three configurations you will find the view as seen with your telescope. There will be no need to flip the atlas upside down or to try and mentally reverse a map in your mind.

The atlas is divided into two parts. The first has general maps of the entire lunar disk and identifies the major lunar seas. Finder charts showing the location of the 60 primary Moon formations are found here.

The second part is made up of 60 separate sections numbered from 1 to 60. The individual charts are subdivided into four pages. The first page is a written description of the formations labeled on the images. On the descriptions of most of the features there will be a number in miles in parentheses next to the name of the crater or formation. This number is its diameter.

The page that follows the descriptions is a digital image of the formations. The next two pages will contain two additional images of the primary digital image. You will immediately notice they are not like the digital photograph. I have converted them to an ink outline image with Adobe Photoshop Elements 4.0. I did not want the atlas to be completely photographic, and the ink outlines give the look and feel of topographical maps. However, the "contour" lines are NOT true contour lines. They are differences in the brightness on the original image. The outlines have somewhat of the wonderful look of the hand drawn images made by many of the great nineteenth-century lunar cartographers, and I believe they enhance the beauty of this atlas.

# Contents

# BACKGROUND

# THE LUNAR SEAS

Let's take a look at the lunar seas before we go into a close-up study of the lunar surface. These seas are visible to the unaided eye and all are easily visible with the simplest of optical aid. Even opera glasses will easily distinguish one sea from another.

At a public star party many people will ask questions like; "What is that dark spot on the Moon in the upper right corner, or what are those large smooth areas?" If they observe through even a small optical finder they will see many areas of dark color. At low magnifications through your scope many persons will ask you what these regions are and what they are called.

The simplest explanation of these dark areas is to say they are great plains of frozen lava. The lava came from asteroids that impacted the lunar surface several billions of years ago. After the impacts molten lava welled up and flooded the impact regions. Just think what a fantastic sight it would have been to see those great crashes. Of course they did not all happen at once, but were spread out over hundreds of millions or even billions of years.

The next three images are of a full Moon in the configurations as previously described in the section "HOW TO USE THE ATLAS". The images are as they would appear in binoculars, finder telescopes or through the telescope at a low magnification of 30× to 60×. Following the last image is the list of the names of these nine major seas and one ocean.

D. Spain, *The Six-Inch Lunar Atlas: A Pocket Field Guide*,
DOI 10.1007/978-0-387-87610-8_1, © Springer Science + Business Media, LLC 2009

Finder chart for the lunar seas as seen with the unaided eye, binoculars, a spotting telescope, or a telescope with a terrestrial eyepiece or and erecting prism.

Finder chart for the lunar sesa as seen through a Newtonian and Cassegrain telescope or a refracting telescope without a diagonal.

**Finder chart for the lunar seas as seen through a refractor or a catadioptric telescope (a Schmidt-Cassegrain or Maksutov) with a diagonal.**

## THE SEAS

1. Mare Crisium, the Sea of Crisis, has a surface area of 77,000 sq. miles. Its floor is dark and is easily seen with the unaided eye. Be sure to show this prominent and impressive little sea to interested parties. It will be examined in greater detail later in the atlas.

2. Mare Fecunditatis, the Sea of Fertility, is about 130,000 sq. miles in area. South of Crisium, this sea in irregular in shape. Its area is nearly the same as that of Finland.

3. Mare Nectaris, the Sea of Nectar, is 39,000 sq. miles in area. This small round sea is very nearly the size of Iceland.

4. Mare Tranquillitatis, the Sea of Tranquility, is comparable to the Black Sea, with 169,000 sq. miles of surface. This roughly circular sea will forever be famous as the first place that humans set foot on another world.

5. Mare Serenitatis, the Sea of Serenity, is a very circular shaped sea with an area of 123,000 sq. miles.

6. Mare Frigoris, the Sea of Cold, is an elongated sea is more like a great wide river as it winds along near the north polar area of the Moon. It has an area of 170,000 sq. miles, which is similar to Sweden.

7. Mare Imbrium, the Sea of Rains, is a very circular sea, which is a little smaller than Venezuela and covers an area of 332,000 sq. miles. It is the second largest lava plain on the lunar surface.

8. Mare Nubium, the Sea of Clouds, is roundish in shape, and its northern border is indistinct as it merges with the Ocean of Storms. It has an area of 95,000 sq. miles, about the same size as the United Kingdom.

9. Mare Humorum, the Sea of Moisture, is aptly named. This little sea of 45,000 sq. miles is like a big bay tucked below the Ocean of Storms.

10. Oceanus Procellarum, the Ocean of Storms, is bigger than any of the lunar seas. This area deserves the title of ocean, with an area of 811,000 sq. miles. It is a little smaller than Greenland in area.

When a full or nearly full Moon is in your skies you can point out these seas to your friends and other interested parties. If viewing through a telescope at 60× or less you might want to use a filter to reduce the light intensity. However, in reality the human eye will quickly adjust to the bright image of the Moon, and few persons will find it uncomfortable after observing for several seconds. Use the finder chart that corresponds to the telescope/binoculars you are using. Try to see how many seas you can find with just the unaided eye.

# FINDER CHARTS

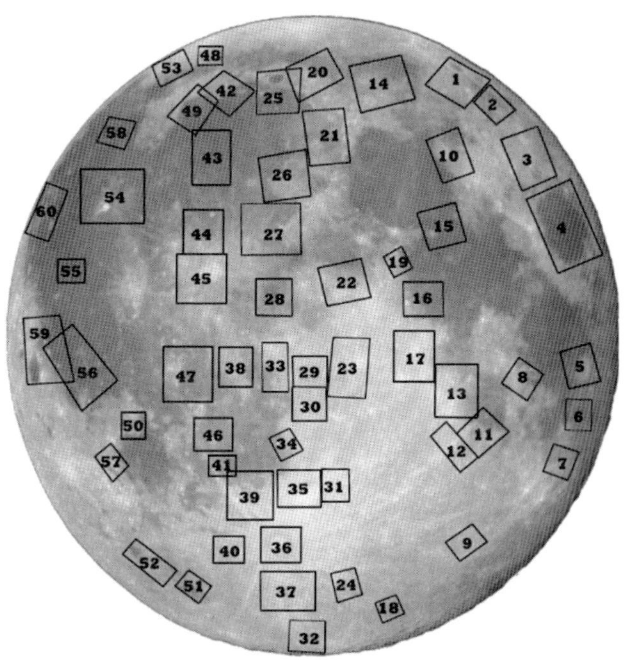


**Finder chart for binoculars, spotting scopes and all telescopes with terestrial eyepeices.**

D. Spain, *The Six-Inch Lunar Atlas: A Pocket Field Guide*,
DOI 10.1007/978-0-387-87610-8_2, © Springer Science+Business Media, LLC 2009

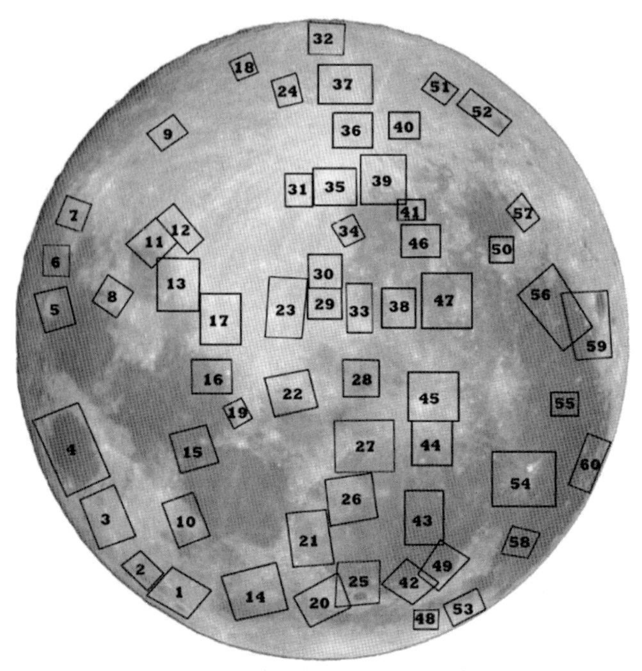


**Finder chart for Newtonians, Cassegrains or refractors without a diagonal.**

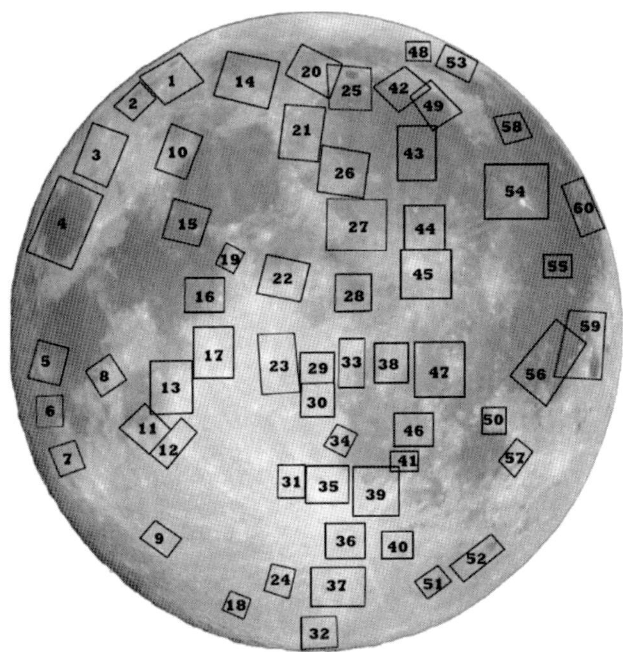


**Finder chart for catadioptrics (Schmidt-Cassegrains, Maksutovs) or refractors with diagonals.**

# THE ATLAS

## CHART 1

# ENDYMION

Endymion (76 miles wide) is a grand crater visible 3 days after the new Moon. The floor is flooded and dark. Its very high walls rise in some places to 2.5 miles above the lava floor. In a 6 inch refractor you will see very little relief on the floor. Catch this great formation on the third day after the new Moon. Be sure to show Endymion to others if you are at a public star party and a thin crescent is hanging in the western sky.

Atlas (53 miles wide) is a great formation and is about 1.7 miles deep. There is a small central peak that is a little off center, perhaps less than half a mile high. The high walls are terraced. Best observed about 4 days after the new Moon.

Hercules (42 miles wide) is almost 2 miles deep. Terraced walls rise above a flat floor with a dark patch in the north. There is a nice 8-mile diameter crater on the southern floor. Observe this impressive formation on the fifth day after the new Moon.

Keldysh (20 miles wide) is located to the north of Atlas. It is a sharp rimmed and deep crater with little detail on the floor.

Cepheus (24 miles wide) has a small central peak. There is a small 8-mile diameter crater on its eastern rim.

If both Atlas and Hercules are visible be sure to point them out at a star party. They make an interesting pair through the eyepiece at a magnification of 120–240.

D. Spain, *The Six-Inch Lunar Atlas: A Pocket Field Guide*,
DOI 10.1007/978-0-387-87610-8_3, 

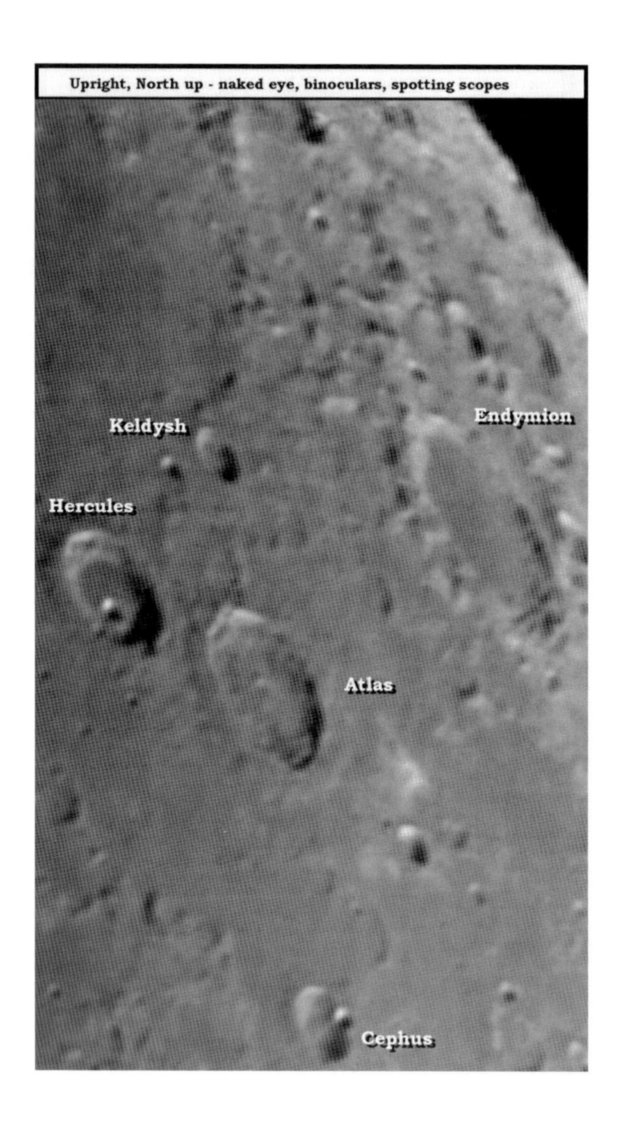
Upright, North up - naked eye, binoculars, spotting scopes
Keldysh
Endymion
Hercules
Atlas
Cephus

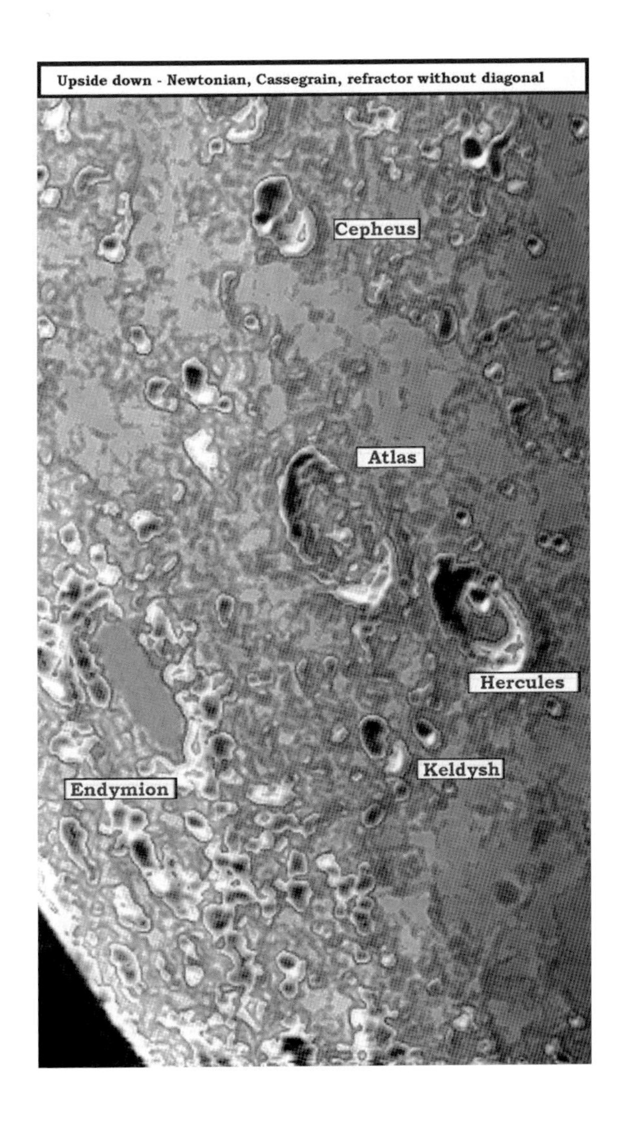
Upside down - Newtonian, Cassegrain, refractor without diagonal
Cepheus
Atlas
Hercules
Keldysh
Endymion

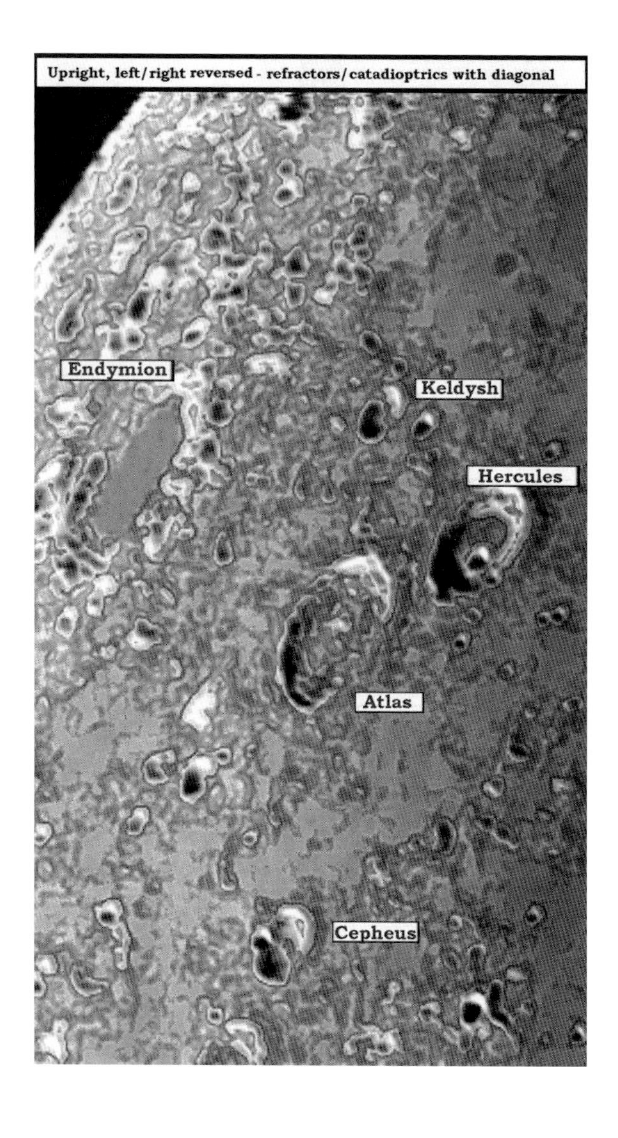
Upright, left/right reversed - refractors/catadioptrics with diagonal
Endymion
Keldysh
Hercules
Atlas
Cepheus

## CHART 2

# MESSALA

Messala (75 miles wide) is a very large formation. The walls are not particularly high for a crater this size; they are highest on the southwest rim and may reach about 2 miles above the floor. The large flat floor is rather rough with small hills, ridges, and craterlets. There are a series of small central peaks that are off-center. There are three noticeable craters on the rim; the most prominent is on the southern edge. Observe this fine crater around the third day after the new Moon. If seeing conditions are good, observe at a magnification of 240. You will see a wealth of detail.

Hooke (22 miles wide) is a crater a little north and west of Messala. The floor is flat, and even at high magnifications it is difficult to see any detail with a 6-inch refractor.

Schumacher (37 miles wide) is directly above and slightly east of Messala. The walls are low, and the formation is battered by younger craters, especially the larger crater that has destroyed most of Schumacher's western rim. The floor is flat, lava-filled, and pretty much featureless.

D. Spain, *The Six-Inch Lunar Atlas: A Pocket Field Guide*,
DOI 10.1007/978-0-387-87610-8_4, © Springer Science+Business Media, LLC 2009

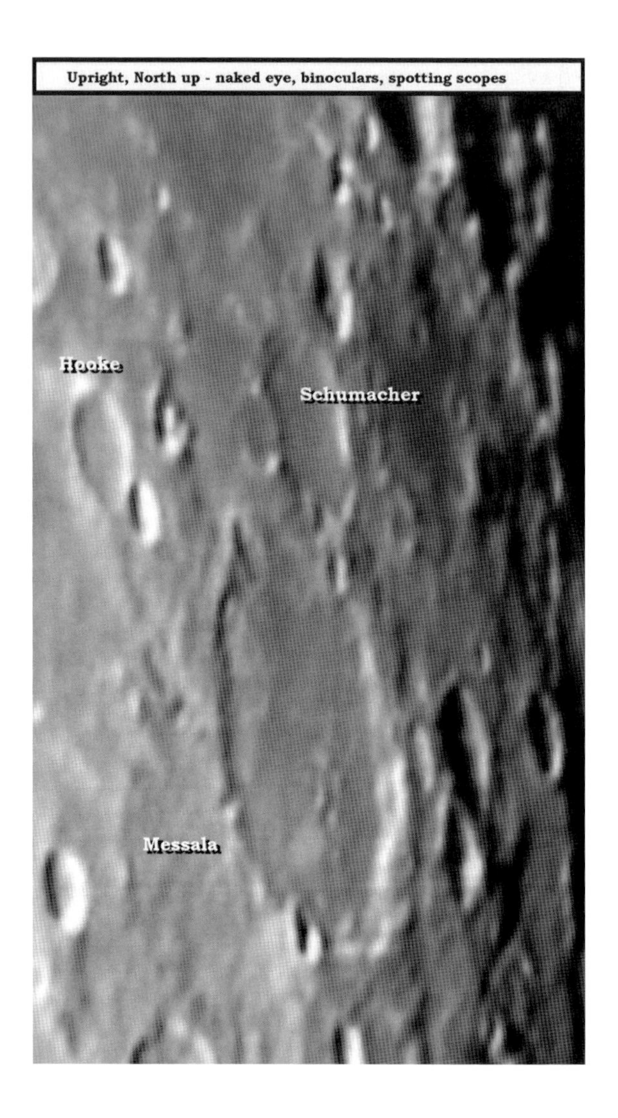
Upright, North up - naked eye, binoculars, spotting scopes
Hooke
Schumacher
Messala

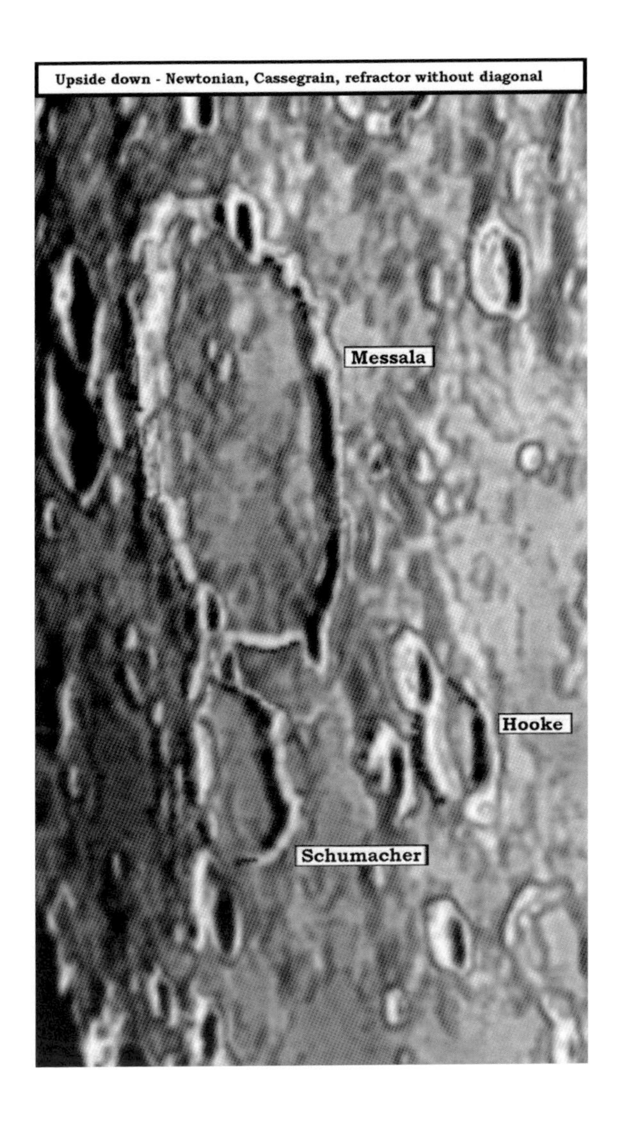
Upside down - Newtonian, Cassegrain, refractor without diagonal
Messala
Hooke
Schumacher

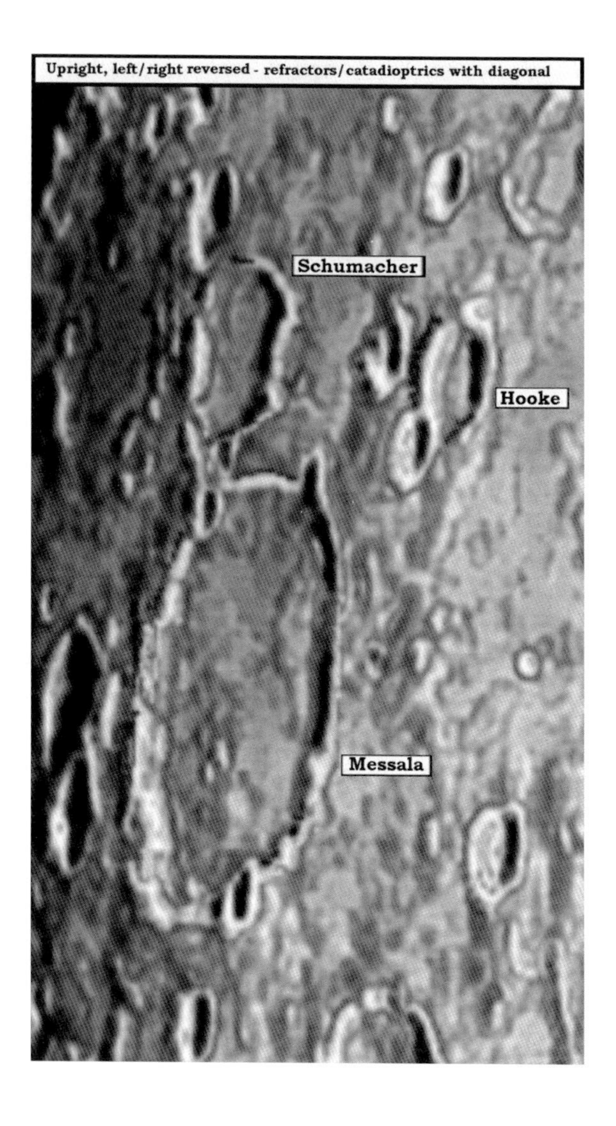
Upright, left/right reversed - refractors/catadioptrics with diagonal
Schumacher
Hooke
Messala

## CHART 3

# CLEOMEDES

Cleomedes (76 miles wide) has highs walls that rise to over 2½ miles in places. The floor is flat and lava filled. There is a small off-center peak that is less than 1 mile high. There are two craters on the floor that are easy to see even in small instruments. While not visible on the photo image there is a cleft on the northern floor that is relatively easy in my 4 inch refractor under grazing sunlight and very easy in the 6 inch. Look for this fascinating formation 3 days after new Moon.

Thalles (27 miles wide) is a crater intruding onto the northwest rim of Cleomedes. Its steep walls rise to almost 2 miles above the floor. There is no central peak, but the floor is rough and has several hills scattered about. It is worth your time to investigate this formation.

Burckhardt (37 miles wide) is just off the northern edge of Cleomedes. Its walls reach above the flat floor to nearly 3 miles. There is a small central peak, less than a mile high. Other than the central peak, the floor shows little detail.

Geminus (52 miles wide) is a prominent crater. The terraced walls are very steep and rise very high above the floor. One source gives a height of 3.1 miles above the floor. I find it hard to argue with that figure. There is a nice central peak that is just under a mile high.

D. Spain, *The Six-Inch Lunar Atlas: A Pocket Field Guide*,
DOI 10.1007/978-0-387-87610-8_5, 

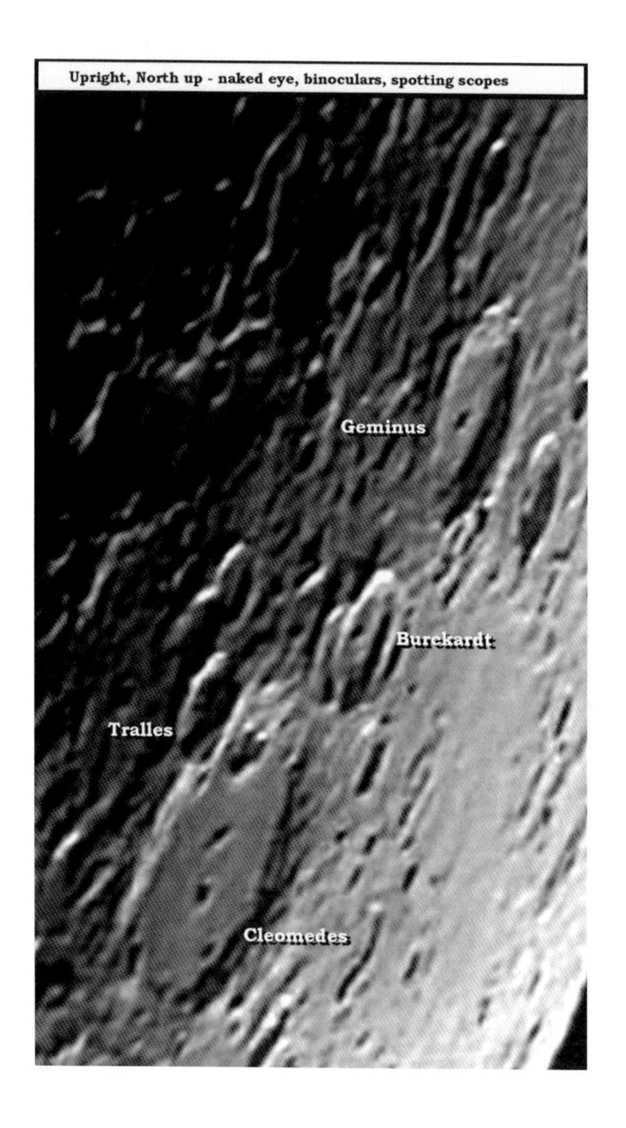
Upright, North up - naked eye, binoculars, spotting scopes
Geminus
Burckardt
Tralles
Cleomedes

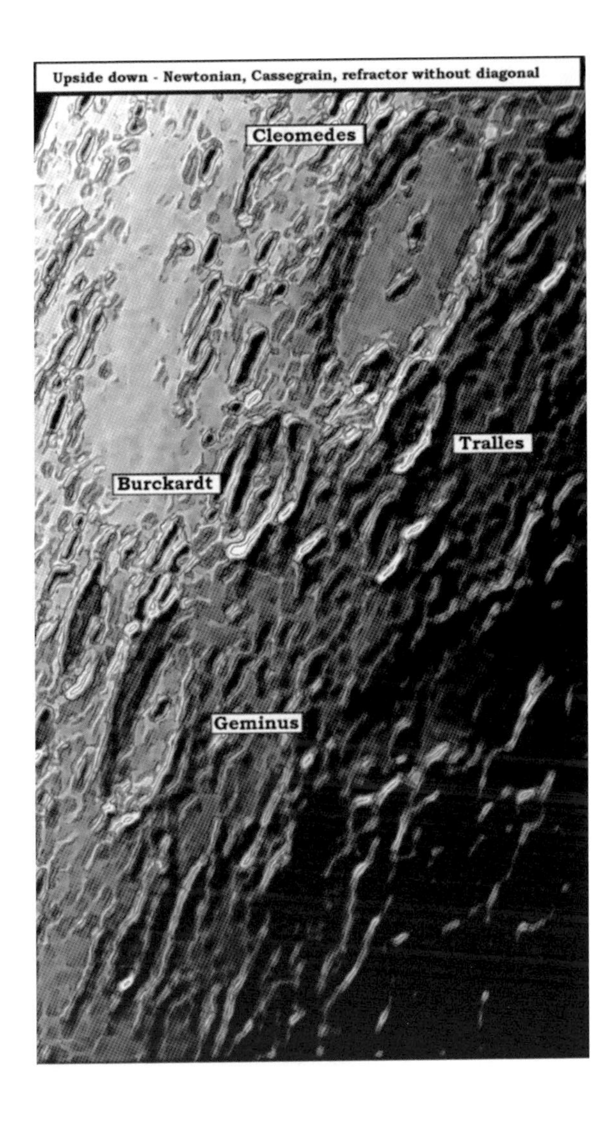
Upside down - Newtonian, Cassegrain, refractor without diagonal
Cleomedes
Tralles
Burckardt
Geminus

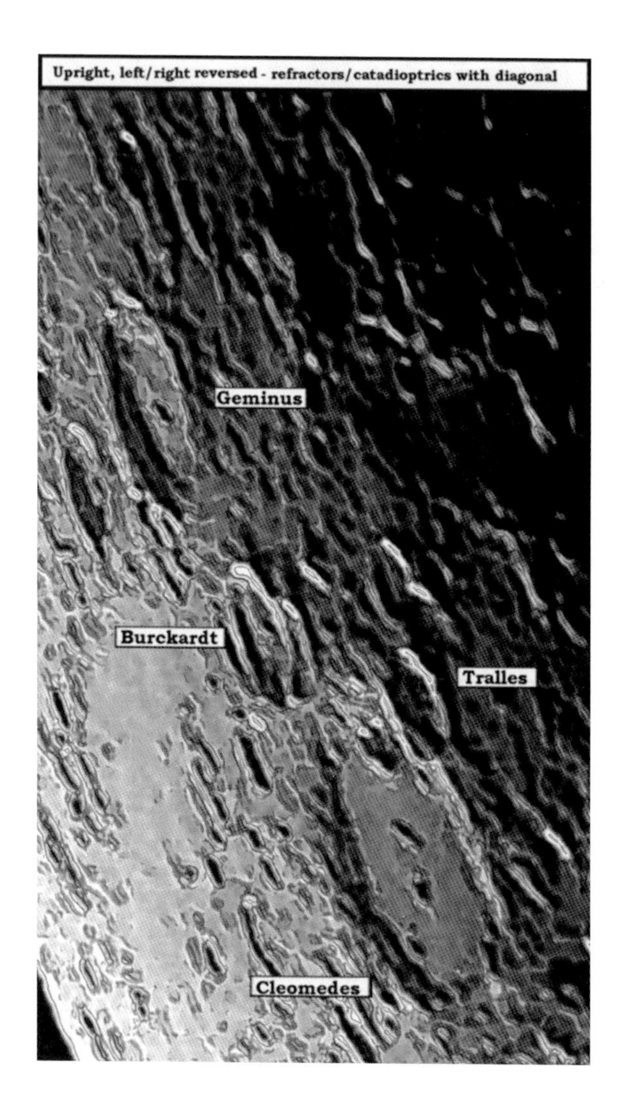
Upright, left/right reversed - refractors/catadioptrics with diagonal
Geminus
Burckardt
Tralles
Cleomedes

## CHART 4

# MARE CRISIUM

Mare Crisium, or the Sea of Crisis, is an exceptional formation. It is actually a lunar basin, a large area of that was flooded by the impact of an asteroid billions of years ago. The flooded lava floor is oval in shape. The east-west direction is 375 miles; the north-south direction is 345 miles. This is the only major mare that is not connected to the great mare system on the earth facing side of the Moon. The floor is surrounded by walls of mountain massifs many thousands of feet high. They rise abruptly from the dark floor. There are many wrinkle ridges, small hills and flooded craters scattered around the sea, concentrating mainly around the edge. At any star party or public observation this is a must formation to show.

Picard (14 miles wide) is a nice crater out in the Crisium Mare. It is over a mile deep and easy to see in a 2.4 inch telescope.

Peirce (11 miles wide) is north of Picard. It is another deep crater for its size, being just over a mile from the top of its rim to the floor.

Swift (7 miles wide) is a small crater just north of Pierce. It is about a mile deep. Under a low sun these three craters remind me of potholes on a highway.

The Sea of Crisis deserves your inspection whenever you have the opportunity to observe it, especially under a rising or setting sun. This entire area is a favorite of mine and I observe it every chance I get. Whether I observe with a 2.4, 4 or my 6 inch refractor I get wonderful views of this amazing sea and all its features. The best time to observe Crisium is about 3 days after new Moon. The best time of the year is spring when the 3 day old Moon will be high in the sky at twilight.

D. Spain, *The Six-Inch Lunar Atlas: A Pocket Field Guide*,
DOI 10.1007/978-0-387-87610-8_6, 

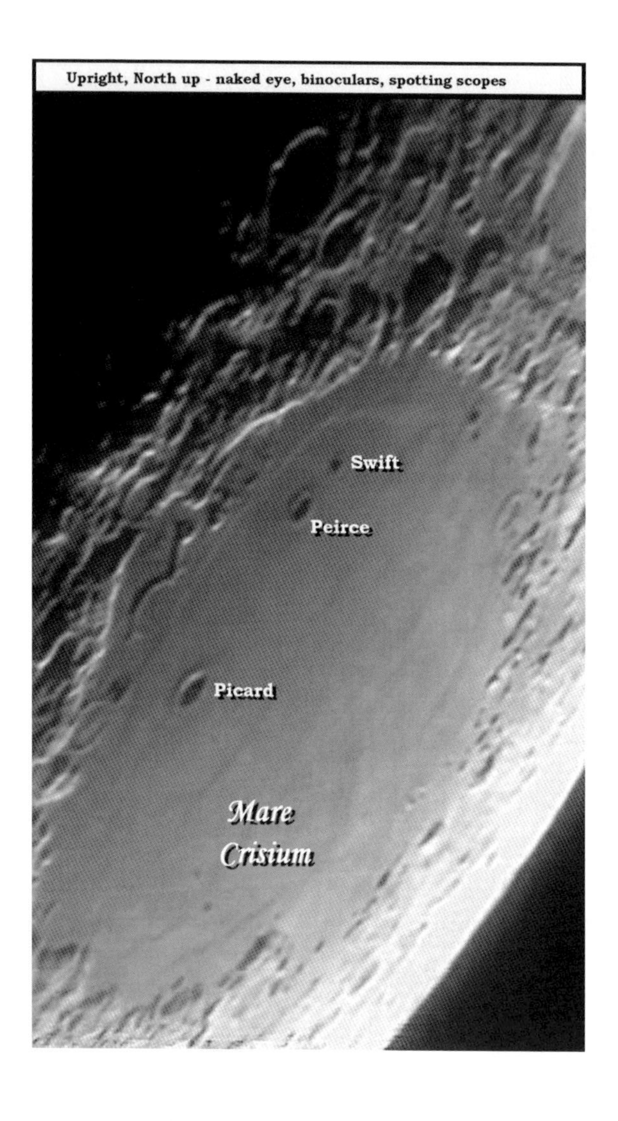
Upright, North up - naked eye, binoculars, spotting scopes
Swift
Peirce
Picard
Mare
Crisium

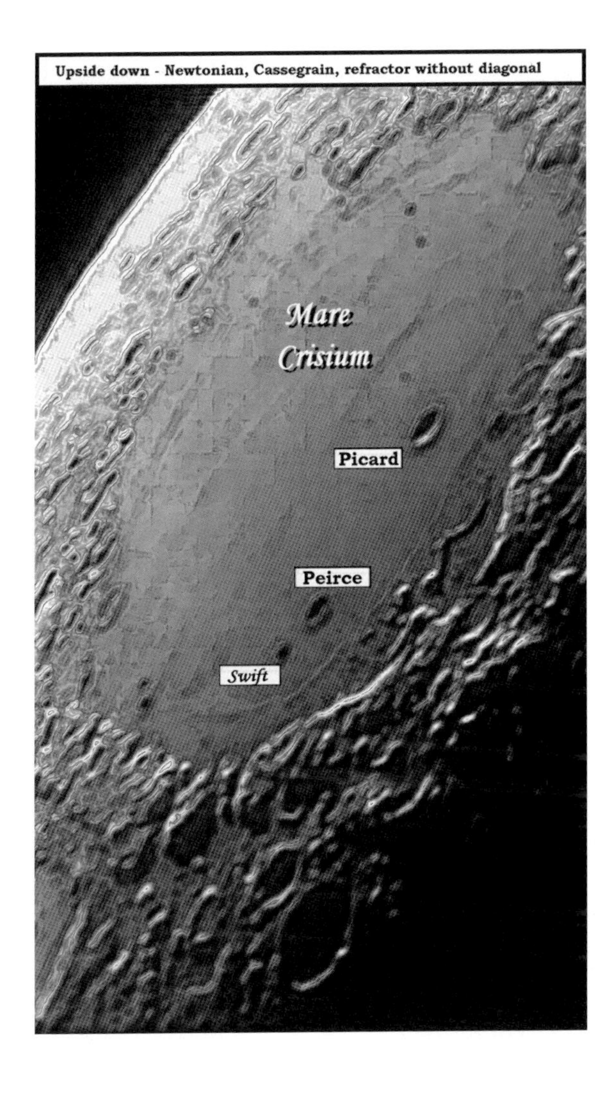
Upside down - Newtonian, Cassegrain, refractor without diagonal
Mare Crisium
Picard
Peirce
Swift

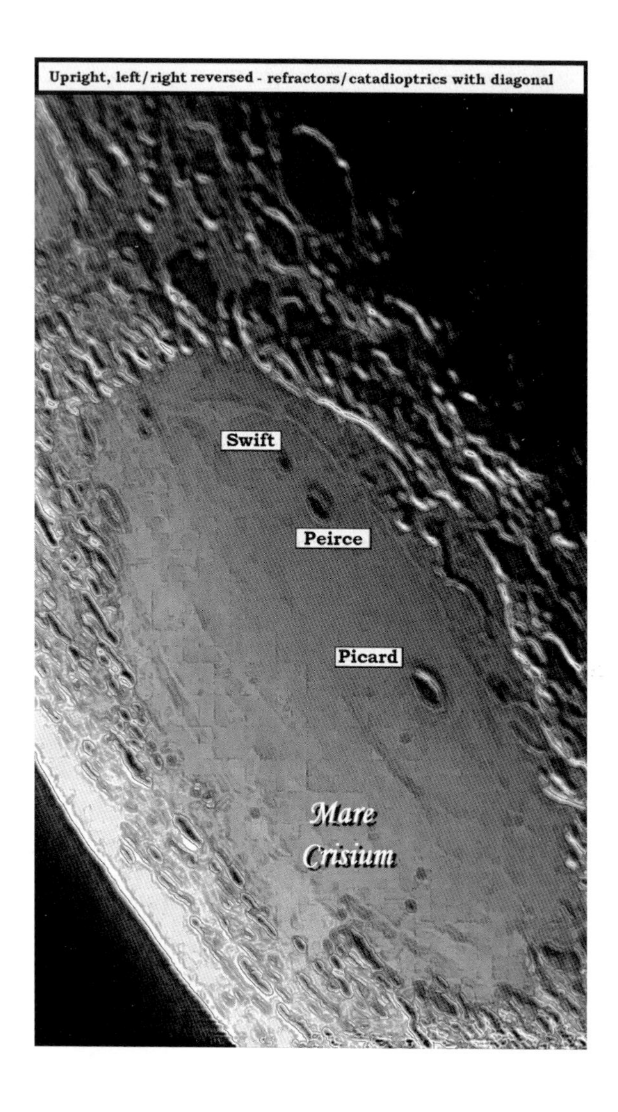
Upright, left/right reversed - refractors/catadioptrics with diagonal
Swift
Peirce
Picard
Mare
Crisium

## CHART 5

# LANGRENUS

Langrenus (80 miles) is a great crater. The walls are terraced and I can see three separate layers of terracing. The walls are probably 2 or more miles high. There is a very nice central mountain mass composed of at least two main peaks. My personal estimation of their height is about 1 mile high. This is a showcase formation at any star party. Observe it 3 days after new Moon.

Barkla (26 miles) is a deep crater to the east of Langrenus. It is less than 2 miles deep. It has a nice central mountain that is under a mile high.

Acosta (8 miles) is a small prominent crater just north of Langrenus and is about a mile deep.

Naonobu (21 miles) is one of three craters north and east of Langrenus. Its rim is about 1½ miles above the floor. I have occasionally glimpsed a small crater on its southern floor

Atwood (18 miles) is over a mile deep. There is a small central peak, which I would guess at less than a ½ mile high

Bilharz (26 miles) is the last of the three craters. It is the largest of the three and the rim is about 1½ miles above the flat and featureless floor.

D. Spain, *The Six-Inch Lunar Atlas: A Pocket Field Guide*,
DOI 10.1007/978-0-387-87610-8_7, 

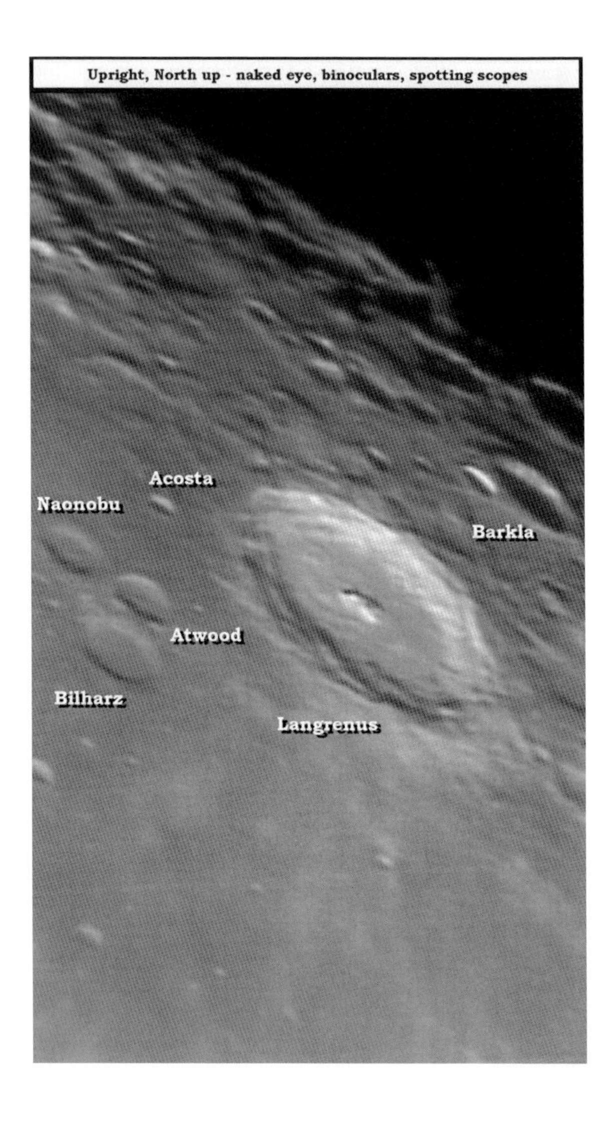
Upright, North up - naked eye, binoculars, spotting scopes
Acosta
Naonobu
Barkla
Atwood
Bilharz
Langrenus

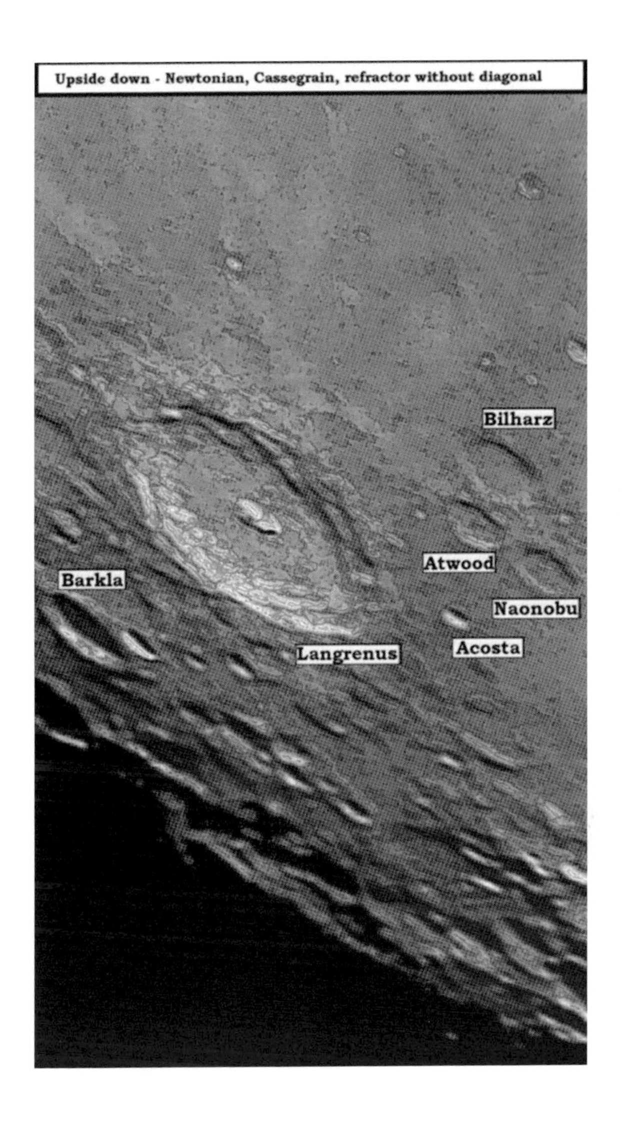
Upside down - Newtonian, Cassegrain, refractor without diagonal
Bilharz
Atwood
Barkla
Naonobu
Acosta
Langrenus

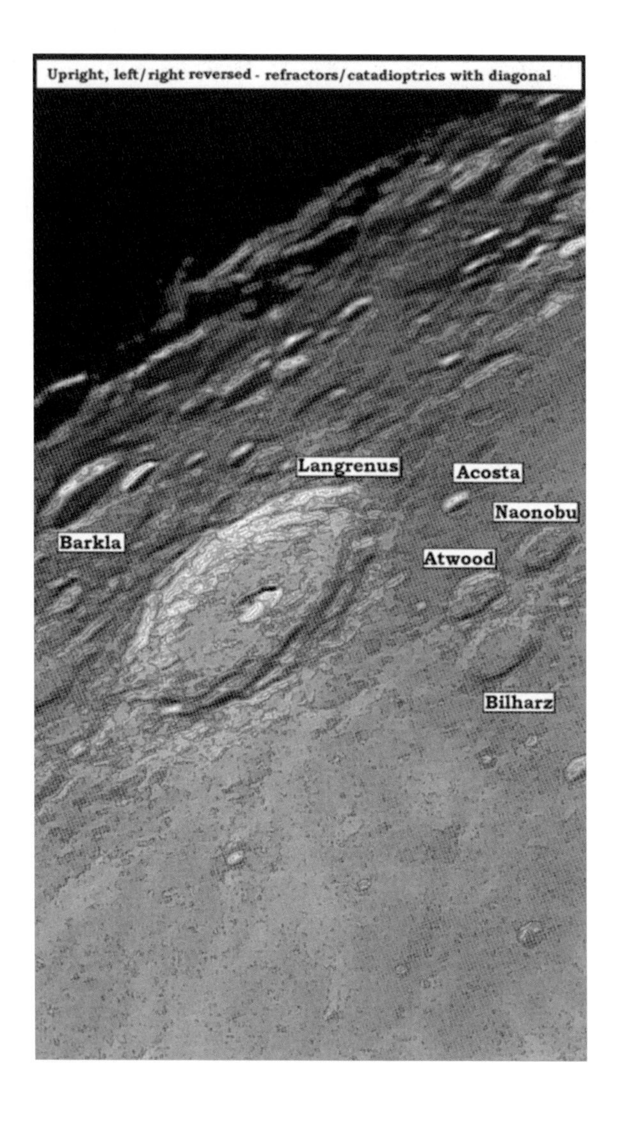
Upright, left/right reversed - refractors/catadioptrics with diagonal
Langrenus
Acosta
Naonobu
Barkla
Atwood
Bilharz

## CHART 6

# VENDELINUS

Vendelinus (90 miles) is a large crater that has suffered from impacts that formed two craters on its northeast and northwest rim. Its low walls are greatly eroded. I doubt if most areas of the rim rises more than 1½ miles above the floor. The floor is flat and many small, and not so small craters, are scattered about the larger ones in the south. There are a couple of small hills on the floor near the southwestern rim that are easy in my 4 inch refractor. While not as grand as many of the formations of this size, Vendelinus is still a fine object to explore.

Lame (50 miles) is a fine crater. Its western rim reaches about 2 miles above the floor. The eastern rim is heavily damaged by other craters and there is a central peak that I estimate to be ½ mile or less in height.

Lohse (25 miles) is west of Lame and half its size. The walls are high; I would guess about 2 miles above the floor, with some terracing. There is a nice central mountain that occupies a great part of the floor and may be a mile high.

Holden (29 miles) is a deep crater, its rim rising over 2 miles above the floor. The walls are terraced and there is a sharp crater on the north rim and another crater on the center of the floor.

This collection of craters will reward the careful lunar explorer with great views around the third day after new Moon.

D. Spain, *The Six-Inch Lunar Atlas: A Pocket Field Guide*,
DOI 10.1007/978-0-387-87610-8_8, 

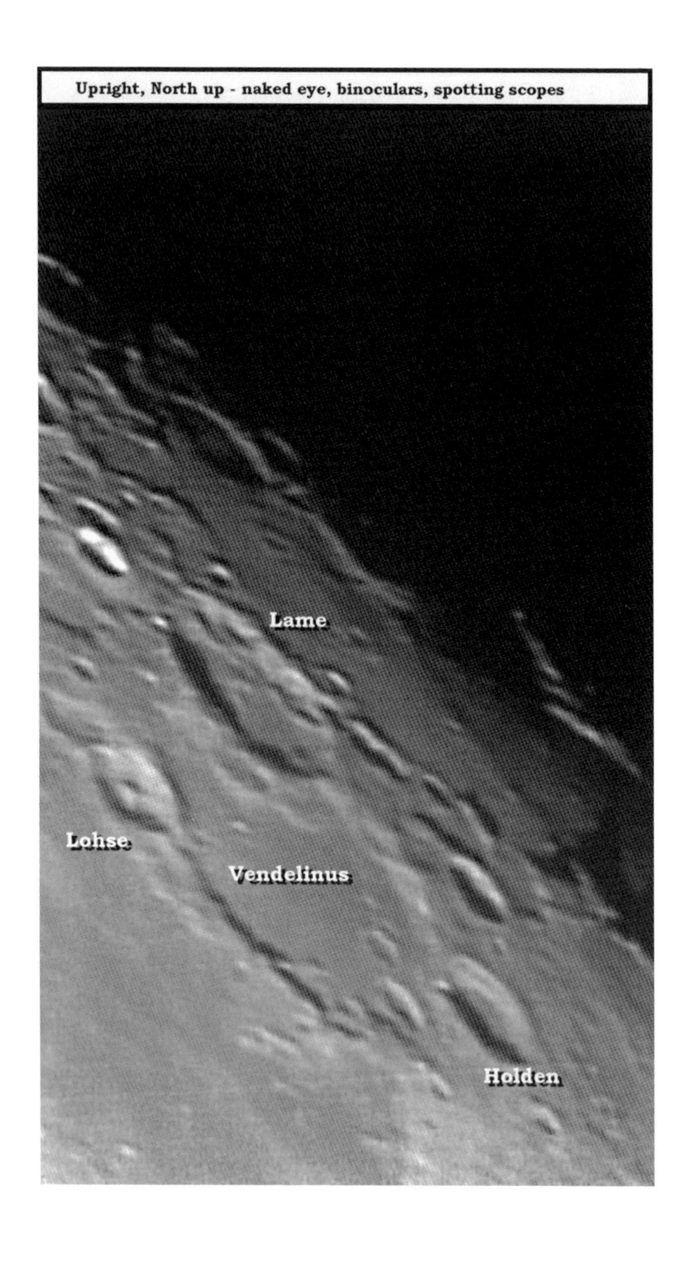
Upright, North up - naked eye, binoculars, spotting scopes
Lame
Lohse
Vendelinus
Holden

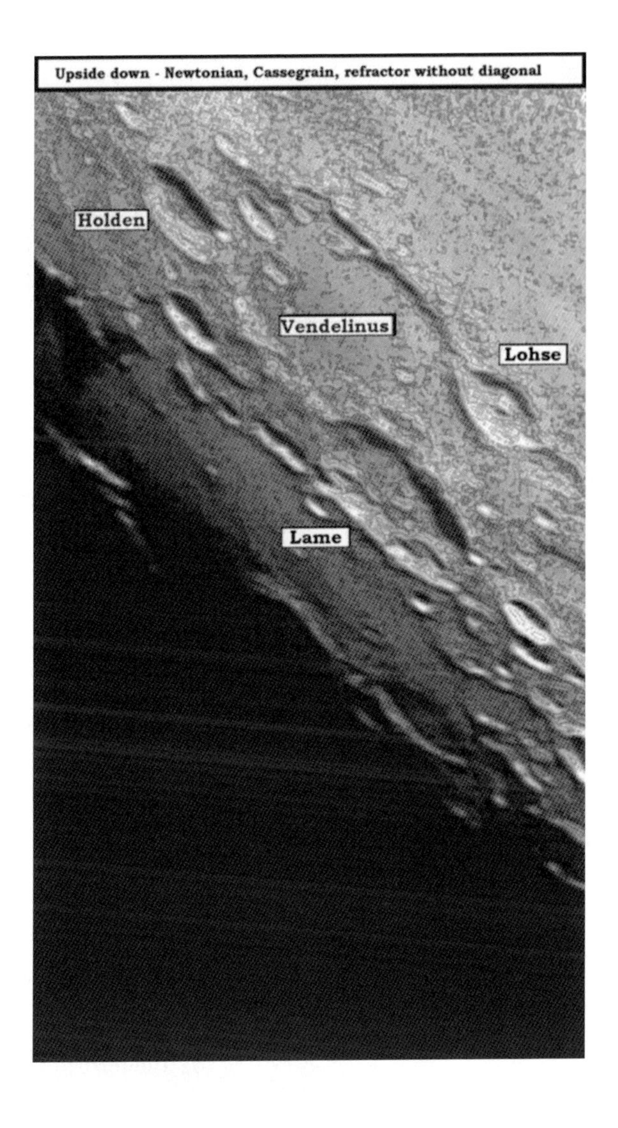
Upside down - Newtonian, Cassegrain, refractor without diagonal
Holden
Vendelinus
Lohse
Lame

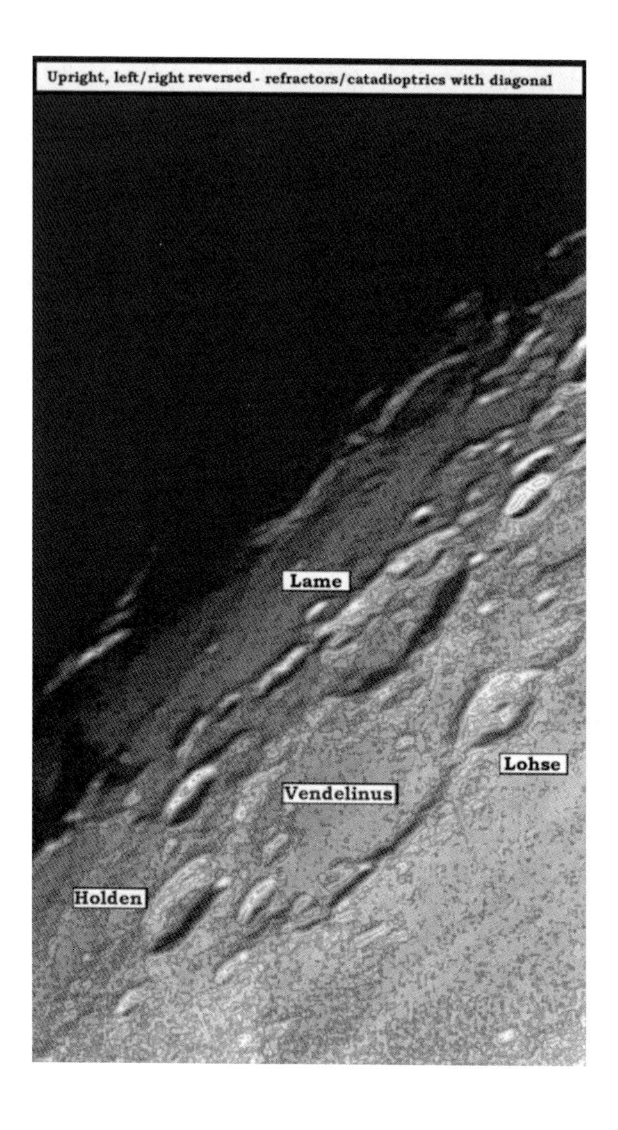
Upright, left/right reversed - refractors/catadioptrics with diagonal
Lame
Lohse
Vendelinus
Holden

## CHART 7

# PETAVIUS

Petavius (110 miles) is a magnificent and exceptional formation. It is one of the lunar giants. The walls reach upwards of 2 miles above the floor and are heavily terraced. There is a central mountain mass that has several peaks and hills. The highest peak reaches up about a mile high. The floor is flat, but there are several rills. The largest and longest runs from the central peaks to the southwestern wall for over 30 miles and is mile wide. It is easily visible in my 2.4 inch refractor. In the 4 inch other rills become visible and the 6 inch reveals even more. Careful observation of the floor will reveal that parts are darker then others, especially in the north. If you are observing with at least 4 inches of aperture, look for a 3 mile crater on the floor to the southeast of the central mountains. If you are at a public observation 3 days after new Moon, be sure to show off this impressive formation.

Wrottesley (35 miles) is a fine crater abutting onto the western wall of Petavius. Its walls are terraced and there is a central mountain. While it is overshadowed by Petavius it deserves some of your observational time.

Hase (50 miles) is a crater adjoining Petavius on its southeastern wall. It is badly battered by other craters and pretty well worn down. The floor is rough and contains an 8 mile crater. Its southern wall has been destroyed by a larger crater.

D. Spain, *The Six-Inch Lunar Atlas: A Pocket Field Guide*,
DOI 10.1007/978-0-387-87610-8_9, © Springer Science+Business Media, LLC 2009

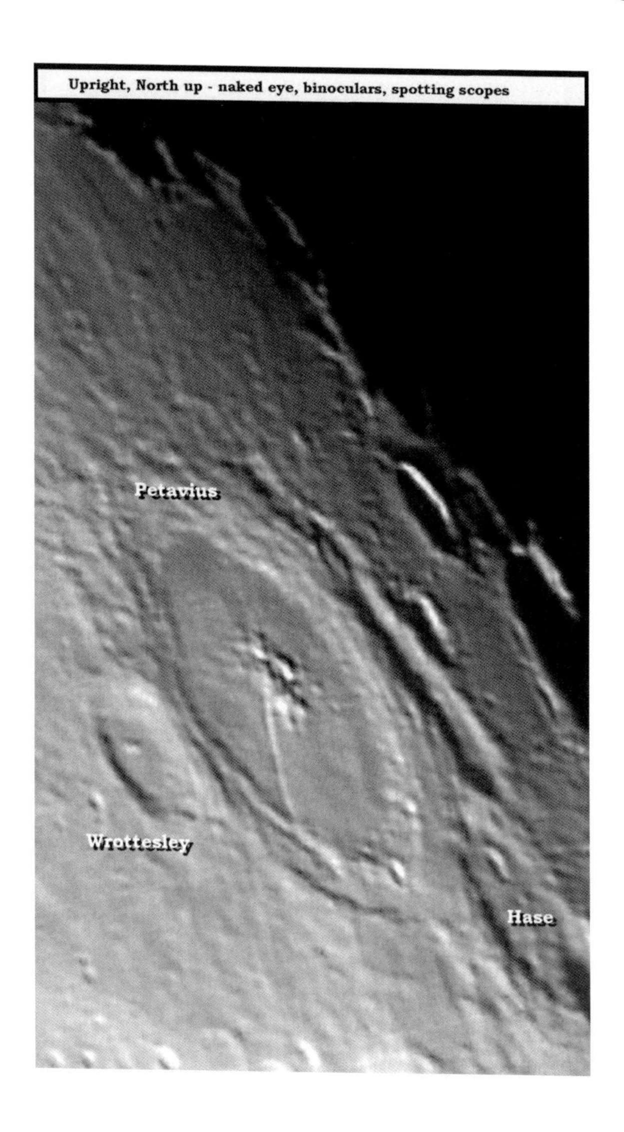
Upright, North up - naked eye, binoculars, spotting scopes
Petavius
Wrottesley
Hase

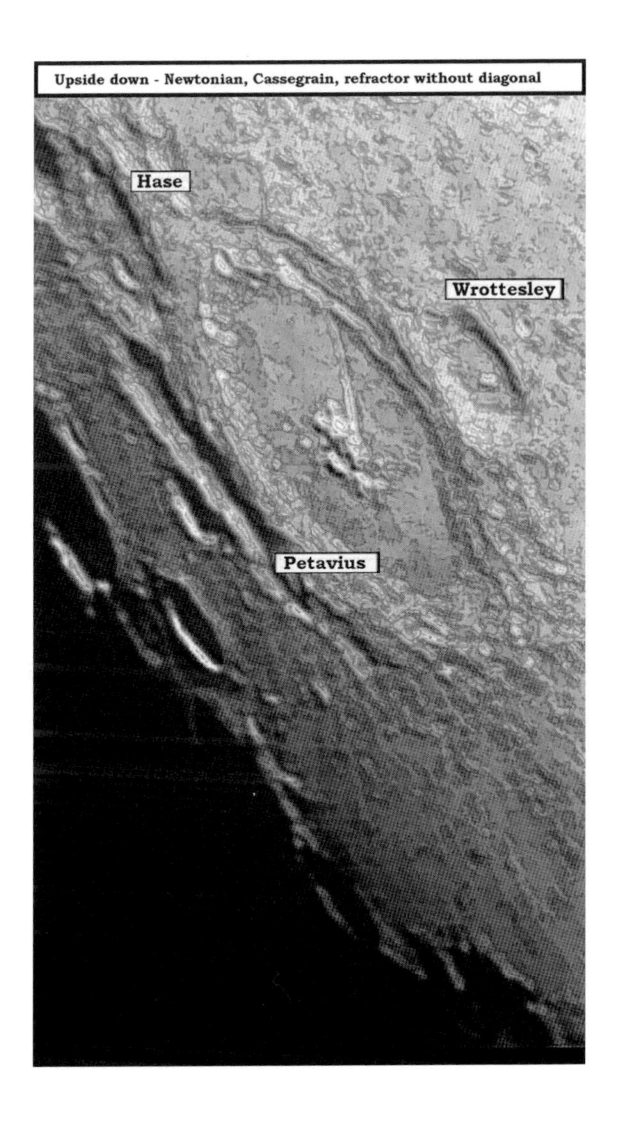
Upside down - Newtonian, Cassegrain, refractor without diagonal
Hase
Wrottesley
Petavius

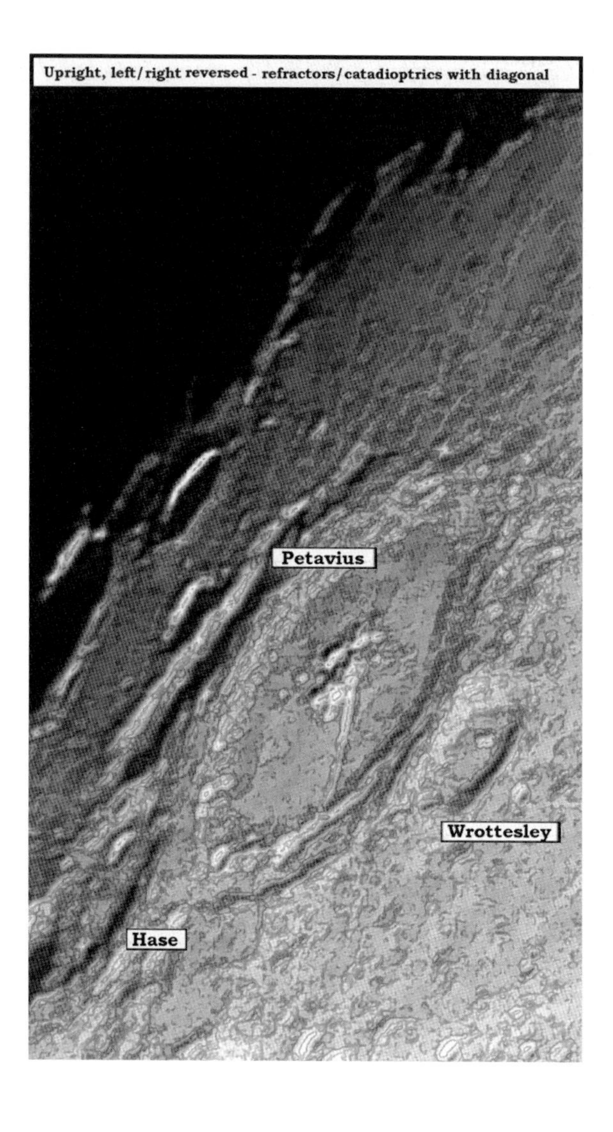
Upright, left/right reversed - refractors/catadioptrics with diagonal
Petavius
Wrottesley
Hase

## CHART 8

# GUTENBERG

Gutenberg (45 miles) is a formation that at first glance looks like an old fashioned keyhole. The western walls rise over a mile above the floor. Part of the eastern rim is destroyed by a later impact and the southern wall is also missing from the impact that makes up the bottom of the keyhole. There are a number of small peaks on the floor. When observing this area with 6 or more inches of aperture you will see a rill running from the east to the northwest. There are also rills outside this crater to the northeast. If this crater is visible at an observation, show it to your viewers for its unusual shape. It will be well placed for observation 5 days after new Moon.

Goclenius (33 miles) has rather low walls that are less than a mile above the floor. Under good seeing conditions my 6 inch refractor reveals a rill that crosses the floor from the southeast to the northwest. Most of the rills around Guttenberg and Goclenius run in this direction.

Magelhaens (25 miles) has walls a little over a mile high. Its featureless floor is flat and flooded with lava.

Colombo (46 miles) is a fine crater to the southeast of Gutenberg. Its walls are over a mile high. There is a small central peak and its inner eastern wall is complex. Look for Colombo 4 days after new Moon

D. Spain, *The Six-Inch Lunar Atlas: A Pocket Field Guide*,
DOI 10.1007/978-0-387-87610-8_10, © Springer Science+Business Media, LLC 2009

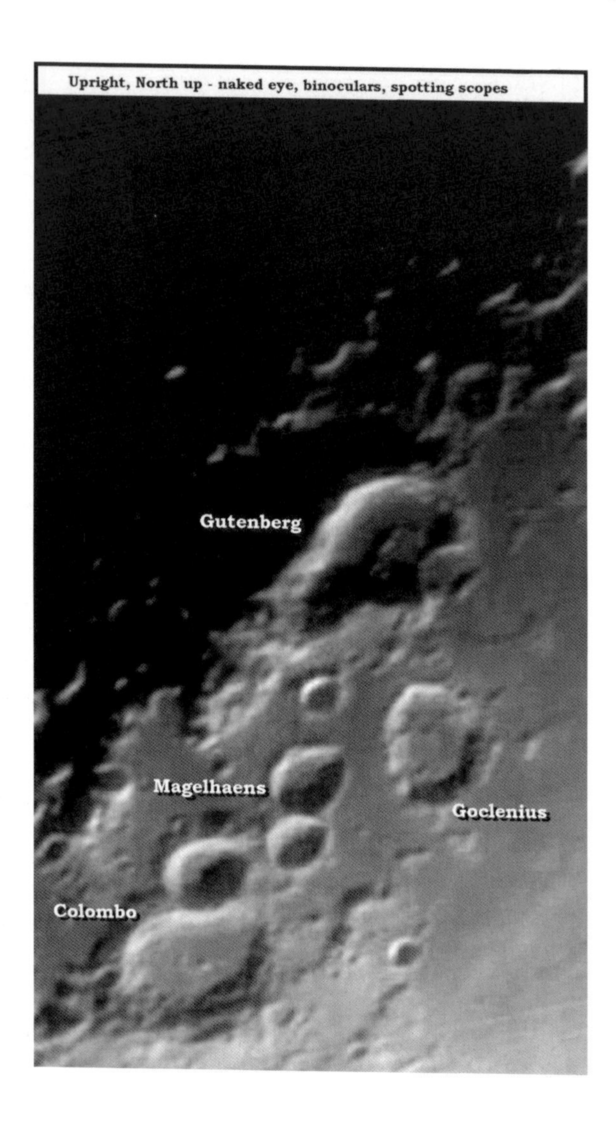
Upright, North up - naked eye, binoculars, spotting scopes
Gutenberg
Magelhaens
Goclenius
Colombo

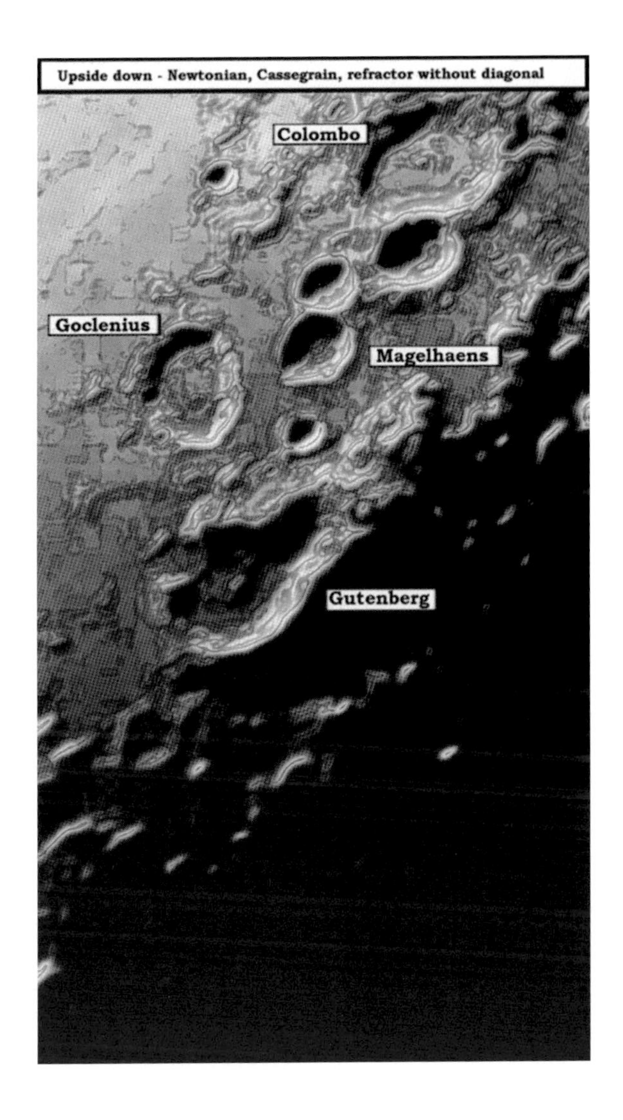
Upside down - Newtonian, Cassegrain, refractor without diagonal
Colombo
Goclenius
Magelhaens
Gutenberg

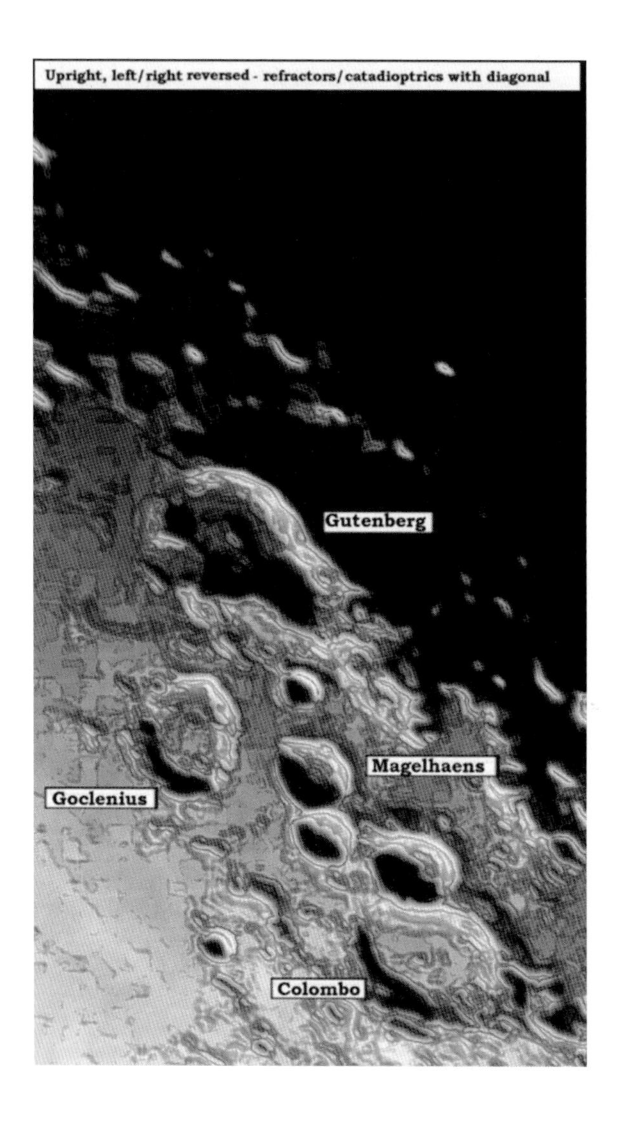
Upright, left/right reversed - refractors/catadioptrics with diagonal
Gutenberg
Magelhaens
Goclenius
Colombo

## CHART 9

# JANSSEN

Janssen (115 miles) is another of the lunar giant craters. The large crater Fabricius is almost entirely within its rather low walls. There are many other craters inside this behemoth. Under grazing sunlight at about 5 days after new Moon, this formation can be seen with 10 power binoculars. There is a low central peak near the center and two great rills. These rills are easily seen on the photo, forming a great arc from Fabricius to Janssen's southeast wall. The northern floor is rough and there is a nice 16 mile diameter crater perched on the northern rim. Though old and battered, under low sunlight Janssen is a must to explore with any size instrument.

Fabricius (47 miles) is a deep crater with a rim that reaches well over a mile above the floor. There is a well formed central mountain mass as well as a small range of mountains to the north. The walls are terraced.

Metius (53 miles) is nearly 2 miles deep and has a group of small rounded hills on the flat floor.

Steinheil (41 miles) has a large flat floor that is almost 2 miles below the rim. I see no detail on the floor, but the walls are terraced.

Watt (40 miles) is almost identical to Steinheil in size and depth. These two craters make a nicely joined pair. Watt's floor is a little rough and its neighbor has obliterated the northern wall. There is a small crater on the western edge of the floor.

D. Spain, *The Six-Inch Lunar Atlas: A Pocket Field Guide*,
DOI 10.1007/978-0-387-87610-8_11, © Springer Science+Business Media, LLC 2009

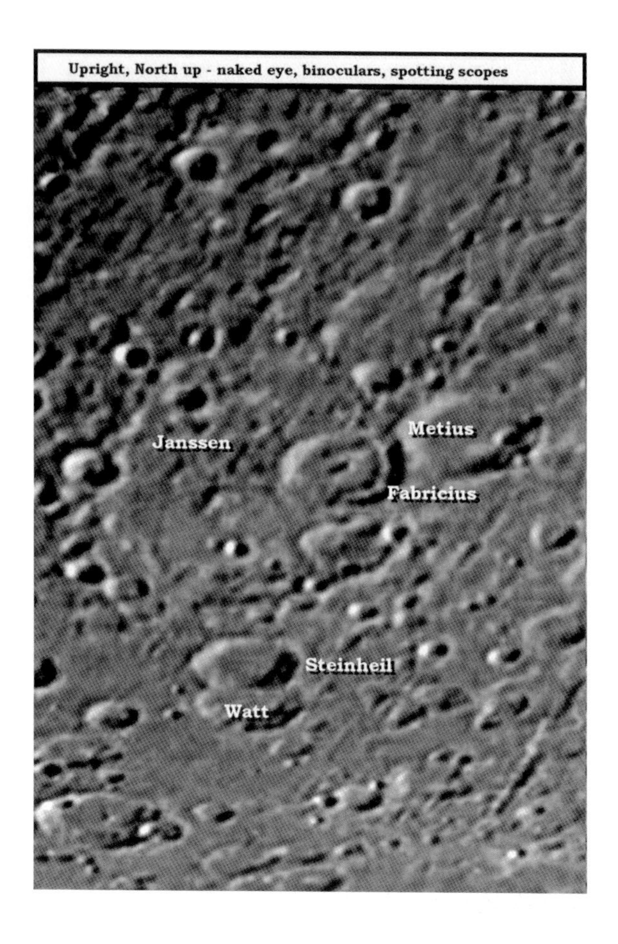
Upright, North up - naked eye, binoculars, spotting scopes
Metius
Janssen
Fabricius
Steinheil
Watt

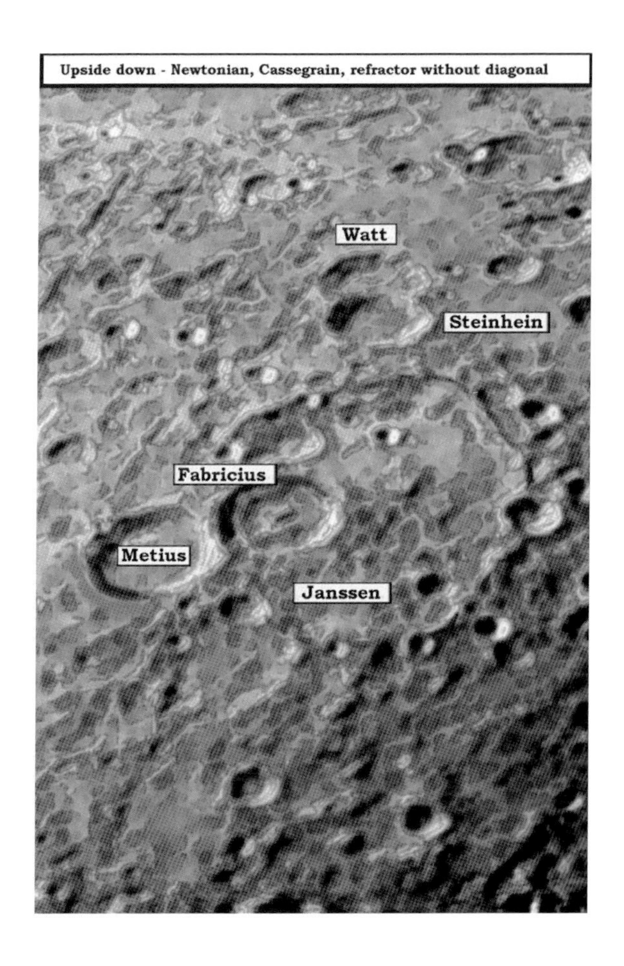
Upside down - Newtonian, Cassegrain, refractor without diagonal
Watt
Steinhein
Fabricius
Metius
Janssen

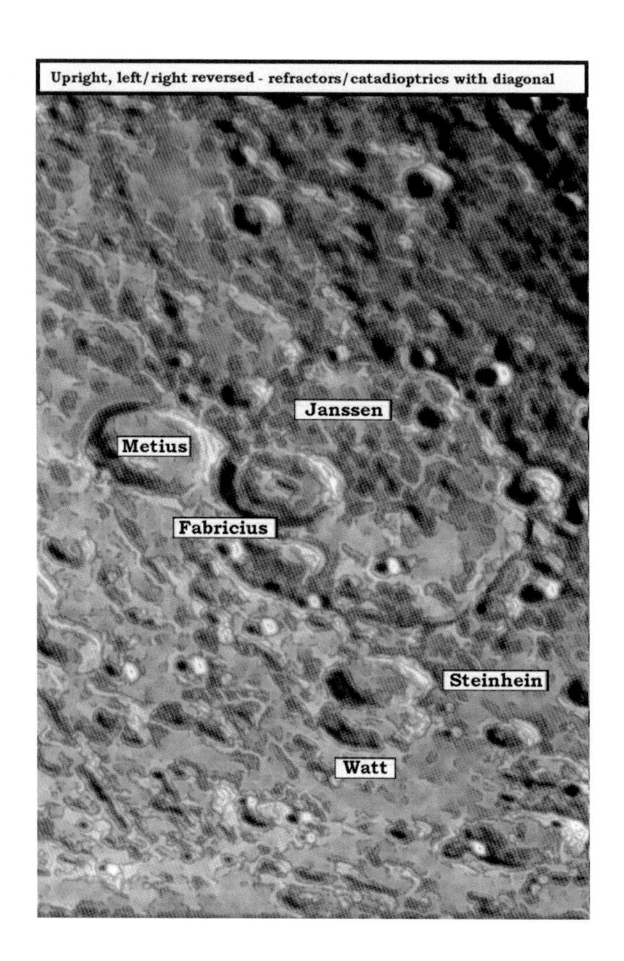
Upright, left/right reversed - refractors/catadioptrics with diagonal
Janssen
Metius
Fabricius
Steinhein
Watt

## CHART 10

# POSIDONIUS

Posidonius (56 miles) is an older formation with rather low walls. There is a nice 6 mile crater off center on the floor and near a small central peak. There is a long ridge, beginning at the southeast and more or less paralleling the eastern rim, curving into the northern floor. In my 6 inch refractor there are several rills visible. The most prominent is the one near the western wall and ends at a ridge. There is a lot of detail on the floor of this old formation. Examine it 5 days after new Moon.

Chacornac (31 miles) is below Posidonius with walls less than a mile high. The floor is rough and there is a little 3 mile crater near the center.

Le Monnier (37 miles) is completely missing its western wall and is really a bay open to the lava sea to its west, the Sea of Serenity. Its floor is dark and parts of the rim are over a mile above Serenity.

G. Bond (19 miles) is a sharp and deep crater. It is about 1½ miles deep. The floor is flat and featureless in my 6 inch refractor.

Hall (24 miles) is really another bay, completely open on its southwestern wall. The higher walls reach to about a ½ mile above the floor. Notice on the photo the rill starting at a small crater above Hall and running southwesterly to G. Bond and below. Under a moderately high sun, this rill is easily visible with 4 inches of aperture

Maury (10 miles) is a sharp rimmed crater. It is almost 2 miles deep and prominent even in small scopes.

Daniell (18 × 14 miles) is an oval shaped crater. Its walls rise to a little over a mile above the flat floor.

D. Spain, *The Six-Inch Lunar Atlas: A Pocket Field Guide*,
DOI 10.1007/978-0-387-87610-8_12, 

Upright, North up - naked eye, binoculars, spotting scopes
Daniell
Maury
Hall
Posidonius
Chacornac
G. Bond
le Monnier

Upside down - Newtonian, Cassegrain, refractor without diagonal
le Monnier
Chacornac
G.Bond
Hall
Posidonius
Maury
Daniell

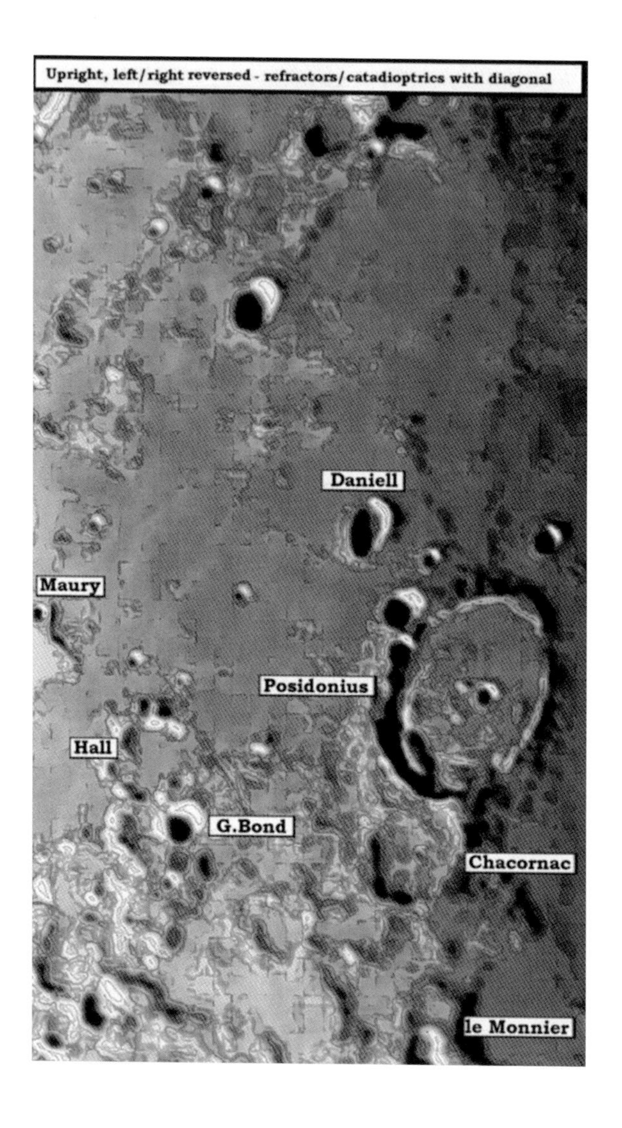
Upright, left/right reversed - refractors/catadioptrics with diagonal
Daniell
Maury
Posidonius
Hall
G.Bond
Chacornac
le Monnier

## CHART 11

# FRACASTORIUS

Fracastorius (75 miles) is the remains of what once must have a great formation. The floor is flooded by the Sea of Nectar, the great lava sea to its north. The rim is very low in the north and indeed part of it is missing completely. The southern walls are much higher, perhaps rising up to a mile or more above the floor. Notice the 16 mile diameter crater intruding on the western rim. Under grazing sunlight, several small craters can be made out in my 6 inch refractor on the floor. There is a tiny central hill and other small hills that can be seen under the low sun. Catch this example of a fine bay 5 days after new Moon. This is a good formation to show to guests at public observations.

Rosse (7 miles) is a small but very prominent crater out in the Sea of Nectar. It is about 1½ miles deep.

Santbech (39 miles) is about 2½ miles deep and the floor is flat and lava filled. There is a little 7 mile crater on the south rim. There is a small peak that is off center.

D. Spain, *The Six-Inch Lunar Atlas: A Pocket Field Guide*,
DOI 10.1007/978-0-387-87610-8_13, 

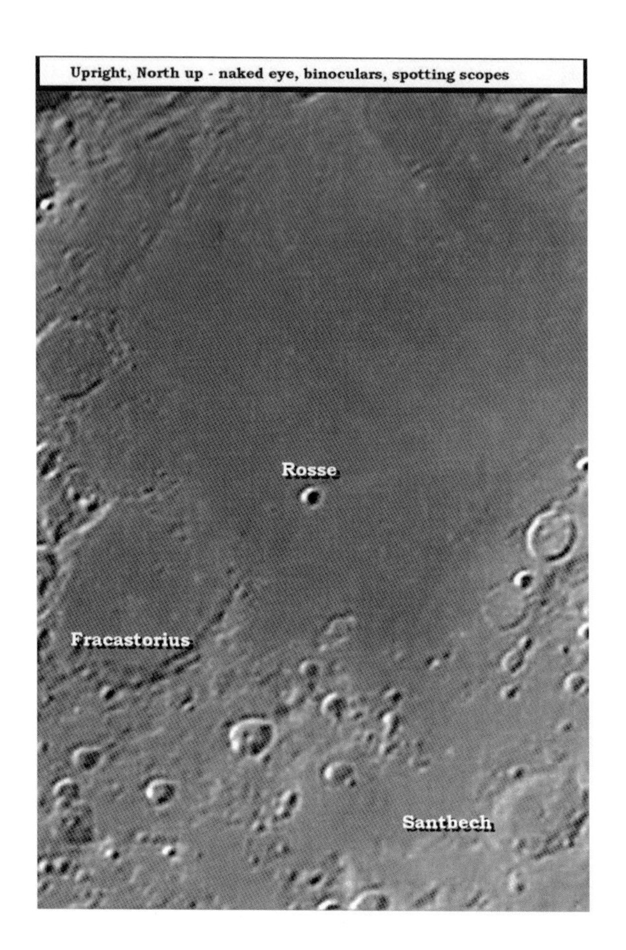
Upright, North up - naked eye, binoculars, spotting scopes
Rosse
Fracastorius
Santbech

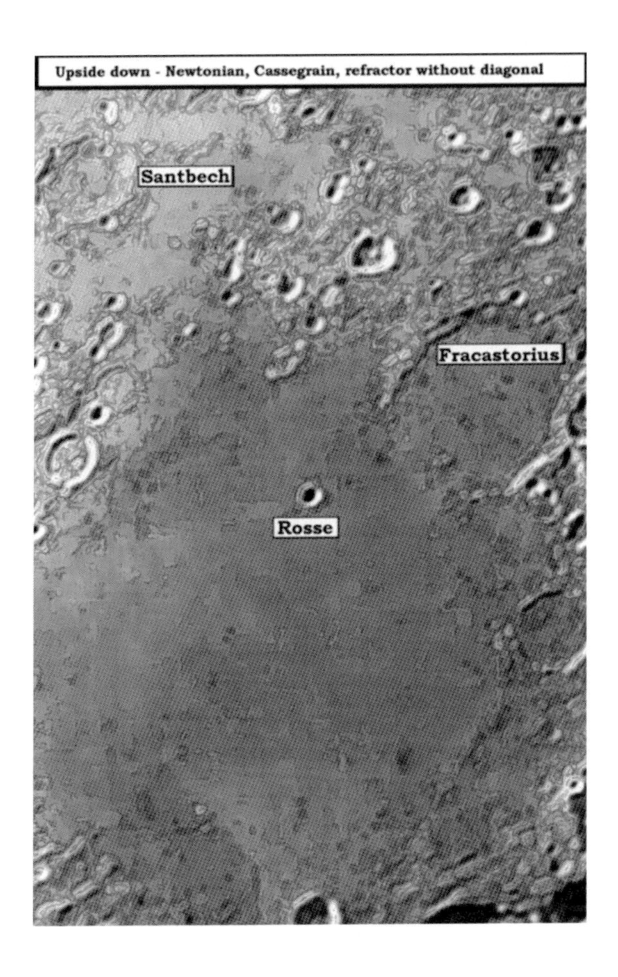
Upside down - Newtonian, Cassegrain, refractor without diagonal
Santbech
Fracastorius
Rosse

Upright, left/right reversed - refractors/catadioptrics with diagonal
Rosse
Fracastorius
Santbech

## CHART 12

# PICCOLOMINI

Piccolomini (53 miles) is an exceptional crater. Its steep walls rise 2½ miles above the floor. The walls are terraced and surround a flat floor. The great central peak rises over a mile and has several summits. Other than the massive central mountain I can see no detail on the floor. This great formation is a grand sight 5 days after new Moon. Be sure to point to this one at a public observation.

Rupes Altai or the Altai Range is long length of curving mountains. It is actually a scarp, or a fault that slopes toward the Sea of Nectar. It is almost 300 miles long, higher at the northern end and much lower at the southern end. With the sun rising on it 5 days after new Moon this formation is an imposing sight.

Polybius (25 miles) is over a mile deep. It has a flat, featureless floor.

Pons (27 by 19 miles) is an irregularly shaped crater. The north and eastern walls are heavily cratered and a good part of the eastern wall is destroyed by a 9 mile crater.

Fermat (24 miles) is a little over a mile deep. Its northern wall is ruined by the intrusion of a 9 mile crater.

Zagut (51 miles) is over a mile deep and the floor has two easily visible craters. The larger is 21 miles across an on the eastern floor. There is a fine 6 mile crater in the center of the floor.

Rabbi Levi (49 miles) is an interesting crater. The western floor is pockmarked by five craters that are easily visible with 4 inches of aperture.

D. Spain, *The Six-Inch Lunar Atlas: A Pocket Field Guide*,
DOI 10.1007/978-0-387-87610-8_14, 

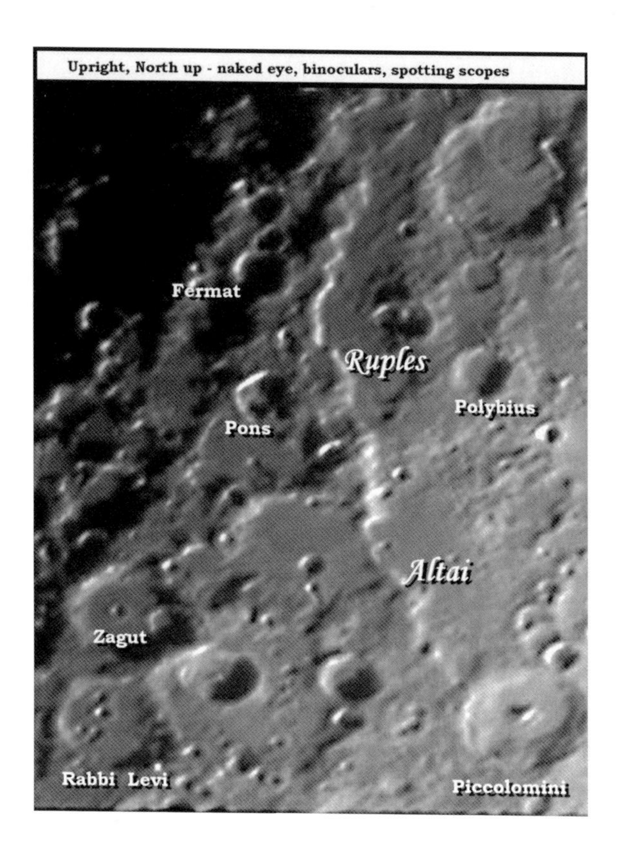
Upright, North up - naked eye, binoculars, spotting scopes
Fermat
Ruples
Polybius
Pons
Altai
Zagut
Rabbi Levi
Piccolomini

Upside down - Newtonian, Cassegrain, refractor without diagonal
Piccolomini
Rabbi Levi
Zagut
Rupes
Pons
Polybius
Altai
Fermat

Upright, left/right reversed - refractors/catadioptrics with diagonal
Rupes
Fermat
Polybius
Pons
Altai
Zagut
Piccolomini
Rabbi Levi

CHART 13

# THEOPHILUS

Theophilus (61 miles), Cyrillus and Catharina form an exceptional trio of craters. The showpiece of these three is Theophilus. Its walls reach up to 2½ miles from the flat floor. This formation is a great showpiece 5 days after new Moon. Its central mountain mass is over a mile high. The great walls are terraced and there is a 5 mile crater in the terraces on the northwest wall.

Cyrillus (59 miles) is to the southwest of Theophilus and its northeastern wall is badly damaged by its neighbor. It is not as deep as Theophilus. There are three off center mountains and a 10 mile crater on the southwestern floor.

Catharina (61 miles) is almost 2 miles deep. There is a large, shallow 27 mile crater on the northwestern area and a 9 mile crater in the south. The floor is filled with detail under a low sun.

Beaumont (32 miles) is a shallow crater and some detail can be seen under a low sun.

Madler (19 miles) is a well defined crater to the east of Theophilus. The walls rise to 1½ miles above the floor.

When viewing the Moon at a magnification of 60, where you can see the whole crescent, the eye is drawn to these three great craters on the western shore of the Sea of Nectar. Absolutely be sure to show this to your guests at a public observation. Use a magnification of 120 to show the details of this region.

D. Spain, *The Six-Inch Lunar Atlas: A Pocket Field Guide*,
DOI 10.1007/978-0-387-87610-8_15, 

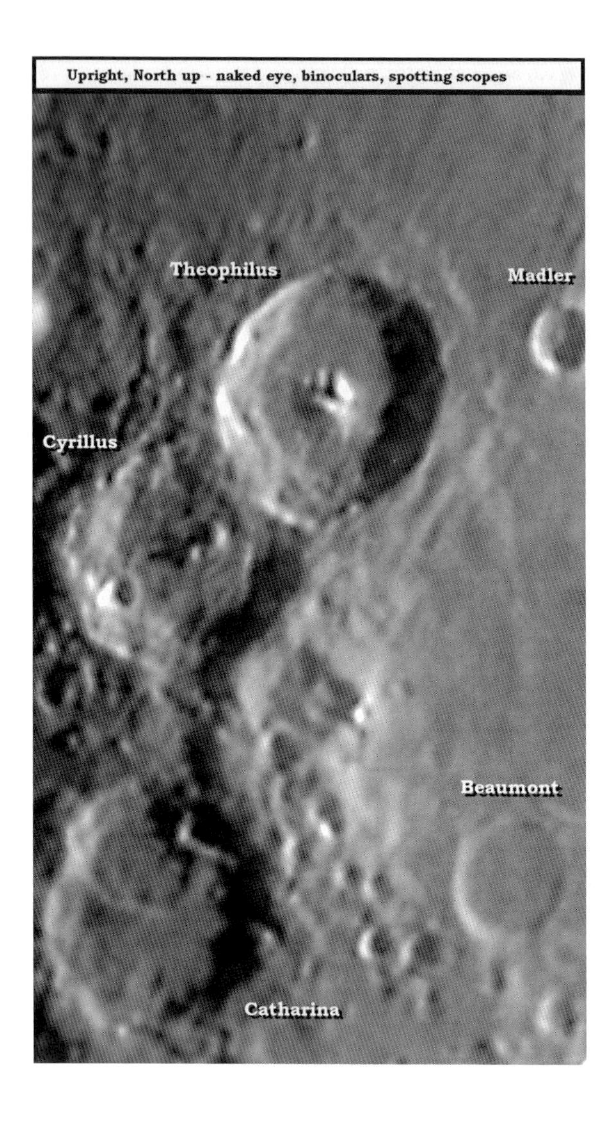
Upright, North up - naked eye, binoculars, spotting scopes
Theophilus
Madler
Cyrillus
Beaumont
Catharina

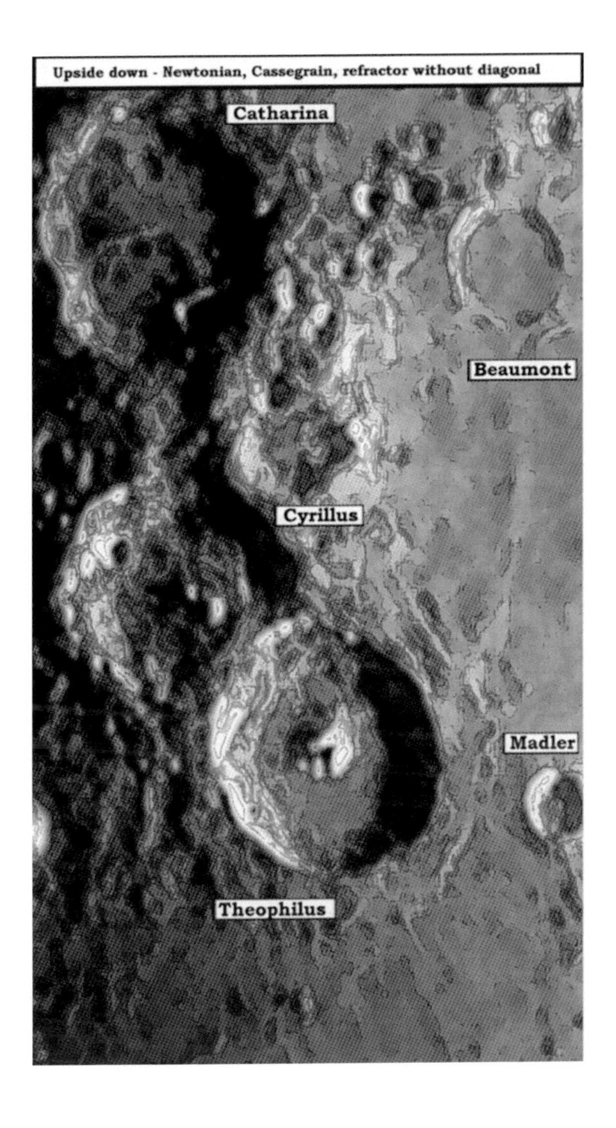
Upside down - Newtonian, Cassegrain, refractor without diagonal
Catharina
Beaumont
Cyrillus
Madler
Theophilus

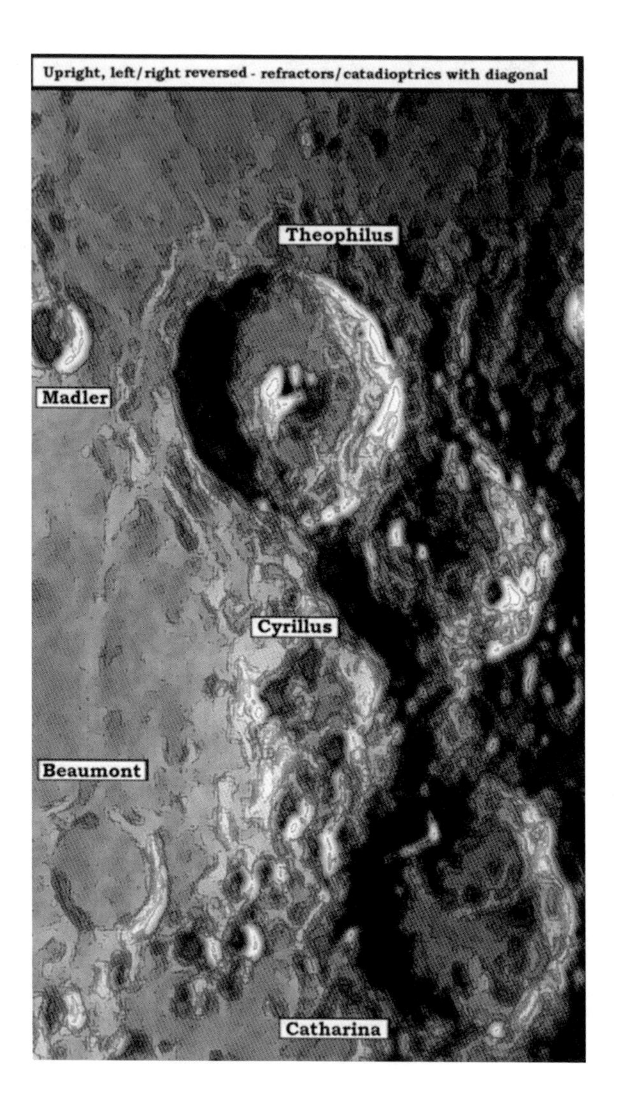
Upright, left/right reversed - refractors/catadioptrics with diagonal
Theophilus
Madler
Cyrillus
Beaumont
Catharina

## CHART 14

# ARISTOTELES

Aristoteles (53 miles) is over 2 miles deep. The walls are extensively terraced and surround the large flat floor. There are two small off center peaks. With Eudoxus about 60 miles to the south these two craters make an attractive pair to show off about 6 days after new Moon.

Eudoxus (60 miles) is a prominent crater about 2 miles deep. The walls are terraced, but not to the extent of its northern neighbor. There is a small central peak and hills on the floor. Both of these craters, and especially Eudoxus, are surrounded by a multitude of hills. Under grazing sunlight the landscape sparkles with sunlight reflecting off the hills.

Mitchell (18 miles) lies on the eastern boarder of Aristoteles. It is about 1½ miles deep. It is one of class of small craters that has been partially destroyed by a larger crater. Normally smaller craters impact on the older, larger formations. Mitchell's western wall has been destroyed by the impact that created Aristoteles.

Galle (13 miles) is a sharp rimmed crater that is over a mile deep. It is easily visible in a 2.4 inch refractor. Find it about 80 miles north of Aristoteles.

Burg (24 miles) is a fine crater that is over a mile deep. There is a nice central mountain on the flat floor.

D. Spain, *The Six-Inch Lunar Atlas: A Pocket Field Guide*,
DOI 10.1007/978-0-387-87610-8_16, 

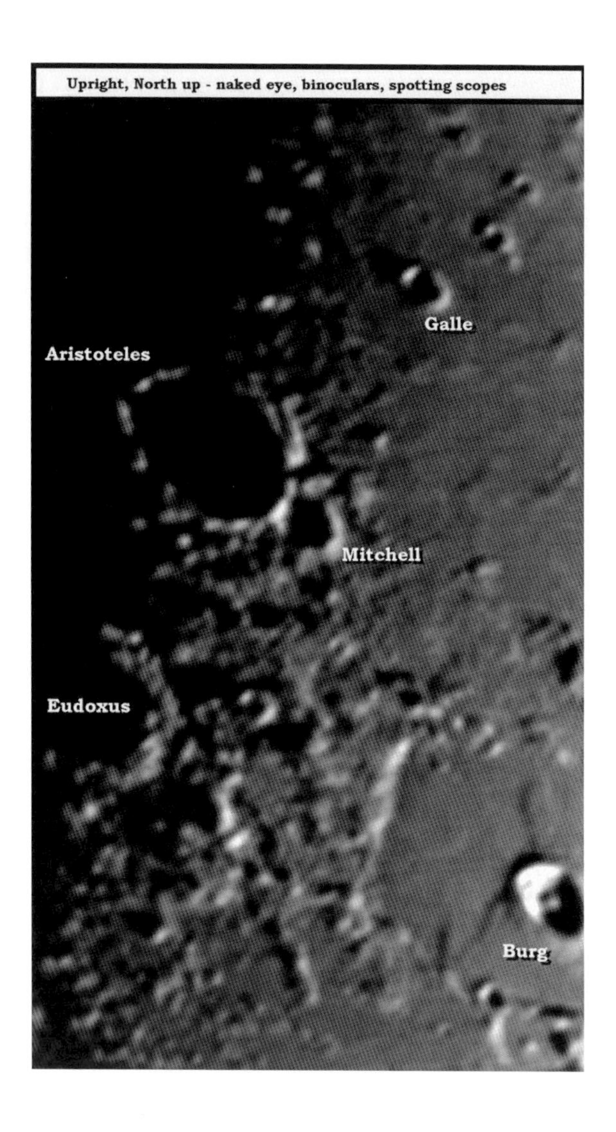
Upright, North up - naked eye, binoculars, spotting scopes
Galle
Aristoteles
Mitchell
Eudoxus
Burg

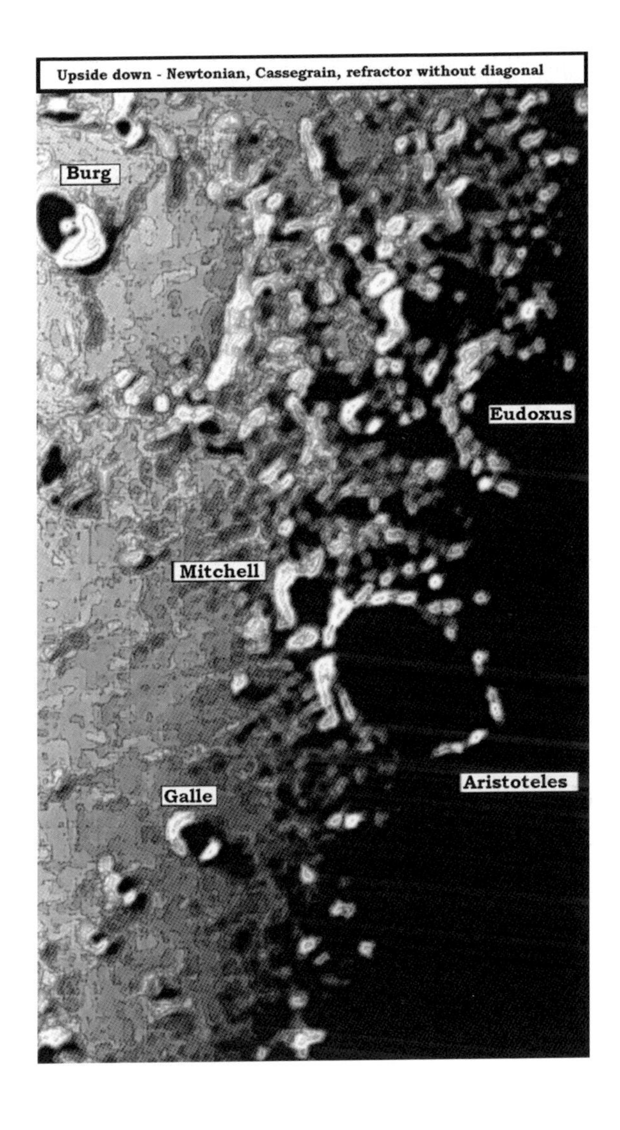
Upside down - Newtonian, Cassegrain, refractor without diagonal
Burg
Eudoxus
Mitchell
Aristoteles
Galle

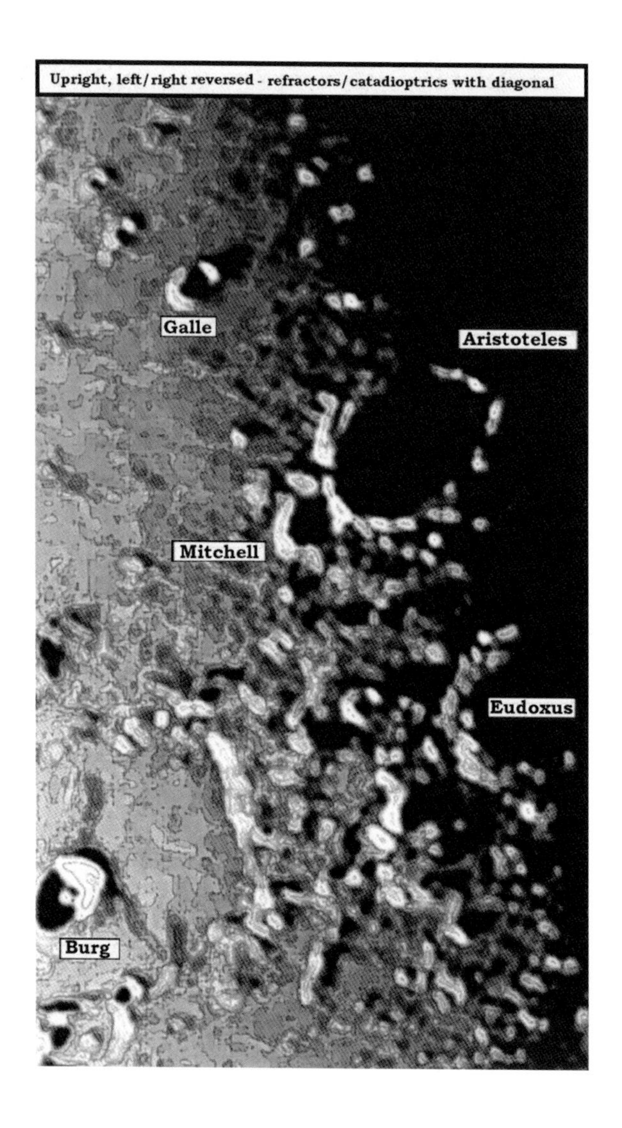
Upright, left/right reversed - refractors/catadioptrics with diagonal
Galle
Aristoteles
Mitchell
Eudoxus
Burg

## CHART 15

# PLINIUS

Plinius (28 miles) is a small, but very prominent crater. The steep walls descend to over a mile to the floor. Some terracing can be seen on the walls and there is a small central mountain, less than ½ mile high. North of Plinius are three rills, one of which shows well on the photo. Also notice the dark streak above Plinius. Observe this area 5 days after new Moon.

Dawes (11 miles) is even smaller than Plinius, but easily stands out, even in my 2.4 inch refractor. It is also about as deep as Plinius and to its northeast.

Ross (16 miles) is just less than a mile deep. It is easy to find with a small scope southwest of Plinius.

Maclear (12 miles) is a flooded crater about 1/3 mile deep. The rim is sharp making this crater easy to identify. The floor is featureless even in my 6 inch refractor at a magnification of 240.

This entire area is very interesting. Look for the multitude of wrinkle ridges, dark and light areas, small craters and isolated peaks on the frozen lava plain.

D. Spain, *The Six-Inch Lunar Atlas: A Pocket Field Guide*,
DOI 10.1007/978-0-387-87610-8_17, © Springer Science+Business Media, LLC 2009

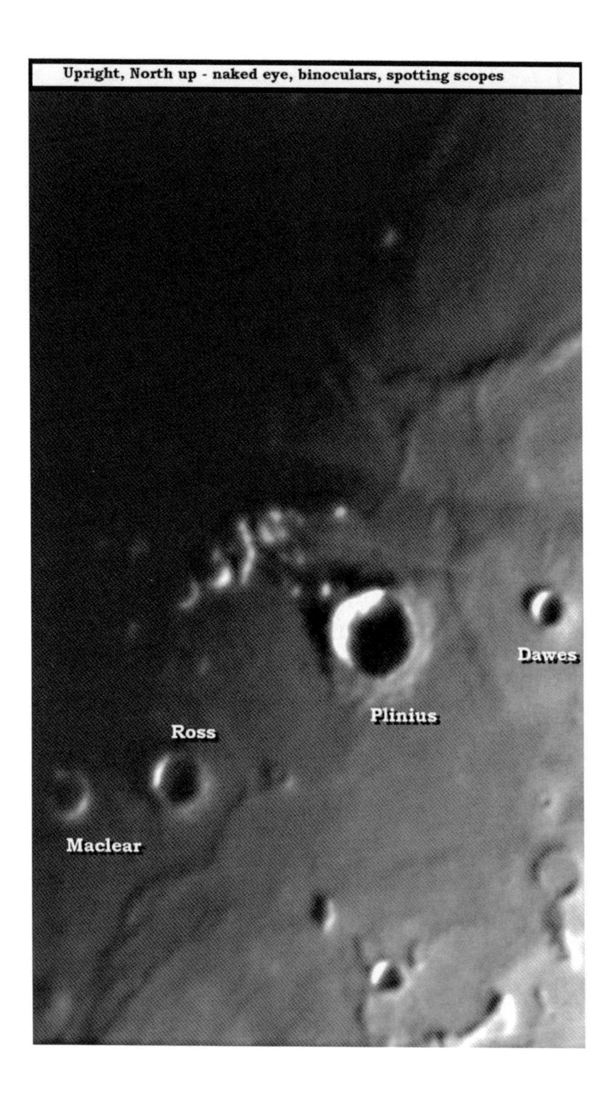
Upright, North up - naked eye, binoculars, spotting scopes
Dawes
Plinius
Ross
Maclear

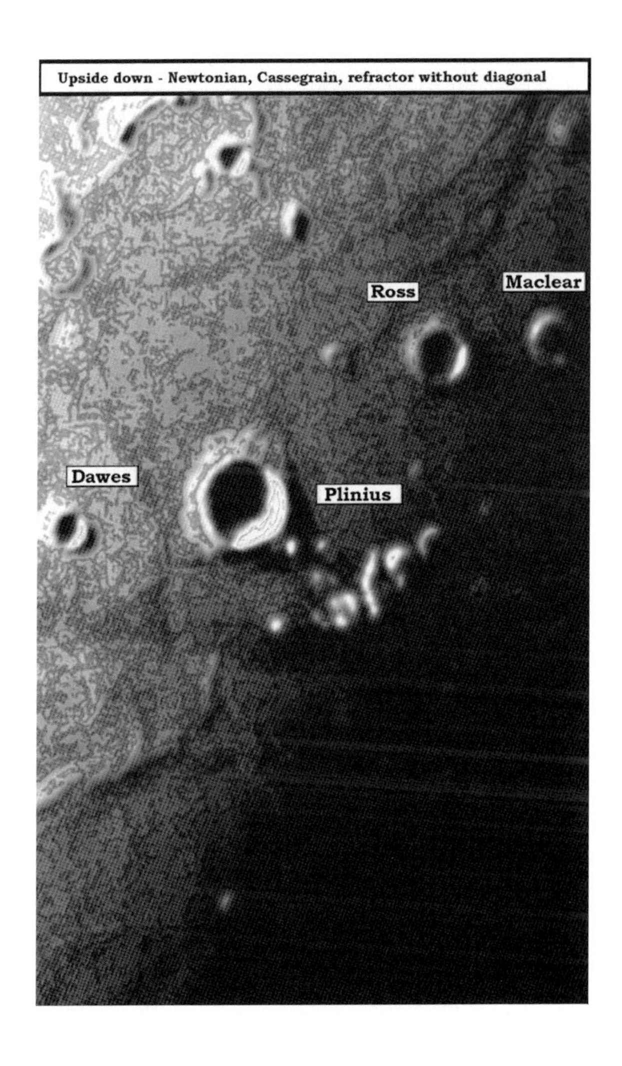
Upside down - Newtonian, Cassegrain, refractor without diagonal
Ross
Maclear
Dawes
Plinius

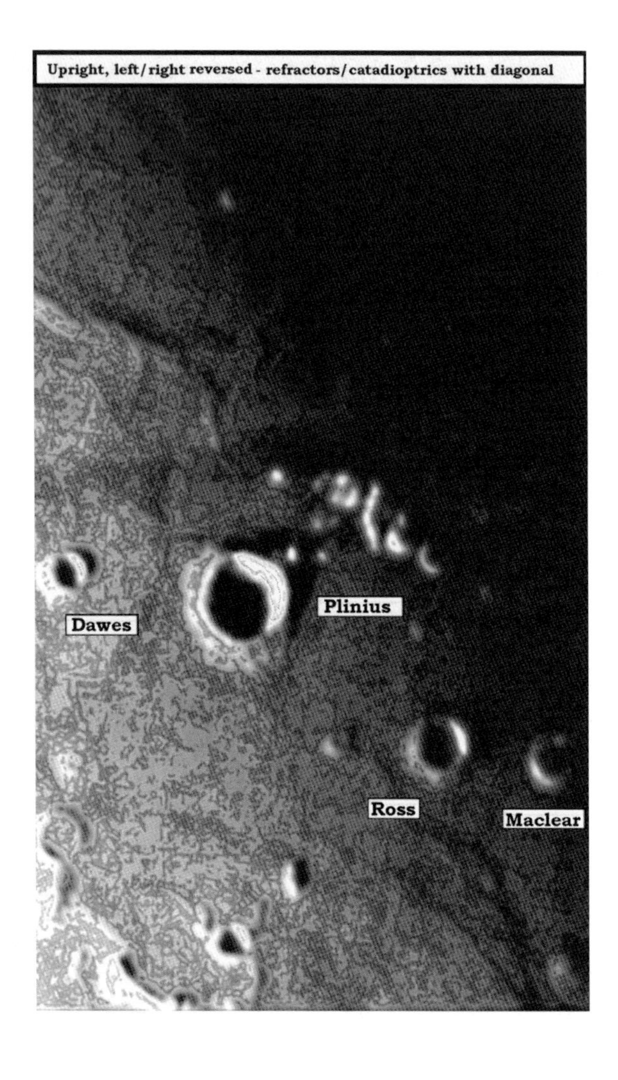
Upright, left/right reversed - refractors/catadioptrics with diagonal
Plinius
Dawes
Ross
Maclear

## CHART 16

# ARAGO

Arago (16 miles) is a sharp rimmed crater about a mile deep. There are several wrinkle ridges to its east and south. On the photo you will see two formations labeled Alpha and Beta. These are large lunar domes that are visible under a low sun. Both are about 12 miles across. I can spot them in my 2.4 inch refractor under grazing sunlight and they are easy with the 4 inch. If they are visible during a public showing be sure to point them out.

Manners (9 miles) is a sharp rimmed crater a mile deep. Look southwest from Arago to find it.

Ritter (19 miles) is another sharp rimmed crater. It is not very deep, just over ½ mile from the rim top to the floor. Under good viewing conditions the floor has a slightly rough appearance.

Sabine (19 miles) is around two or three miles from Ritter and is very similar in appearance. The diameter is almost the same, though it is a little deeper. I see no detail on the floor in my 6 inch refractor.

Apollo 11 Landing Site. I am including this because of the great historical importance that happened here. On July 20, 1969 humankind made its first step off of Earth to another world. I remember it like it was yesterday. I was working the night shift then, but took the evening off. I was not going to miss this for anything. Be sure to point out the landing area to your guests. Of course you will see only a smooth area on the southwestern region of the Sea of Tranquillity, but I almost always have someone ask me where the Eagle landed. The best time to show this historical region is 5 days after new Moon.

D. Spain, *The Six-Inch Lunar Atlas: A Pocket Field Guide*,
DOI 10.1007/978-0-387-87610-8_18, 

Upright, North up - naked eye, binoculars, spotting scopes
Alpha
Beta
Arago
Manners
Ritter
Sabine
APOLLO 11
LANDING SITE
JULY 20, 1969

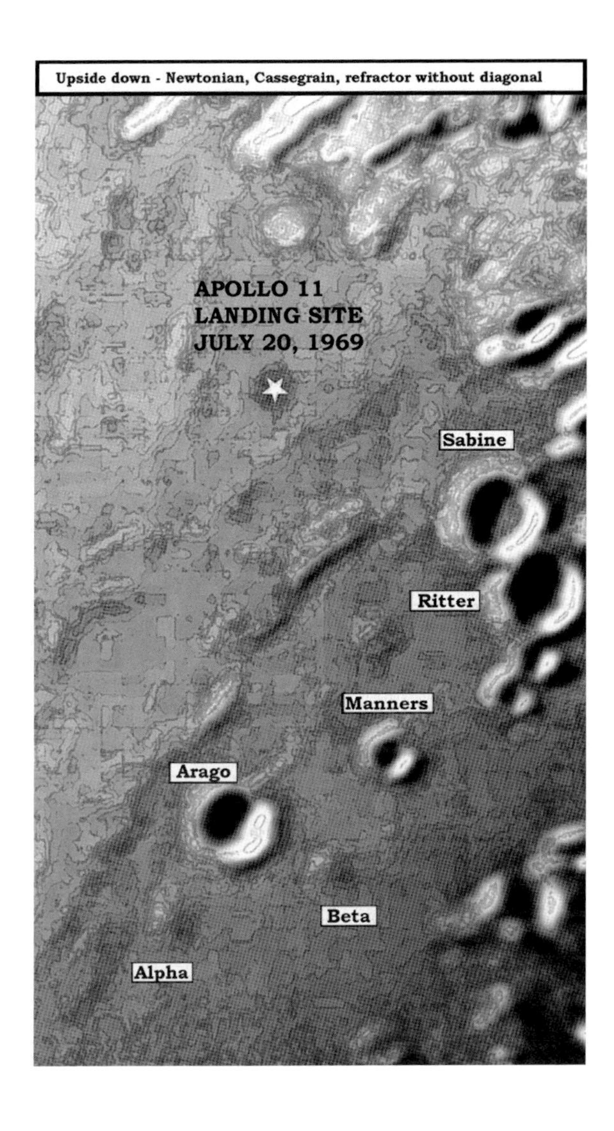
Upside down - Newtonian, Cassegrain, refractor without diagonal
APOLLO 11
LANDING SITE
JULY 20, 1969
Sabine
Ritter
Manners
Arago
Beta
Alpha

Upright, left/right reversed - refractors/catadioptrics with diagonal
Alpha
Beta
Arago
Manners
Ritter
Sabine
APOLLO 11
LANDING SITE
JULY 20, 1969

## CHART 17

# ALFRAGANUS

Alfraganus (13 miles) is a small crater located in a highlands region. It is deep, over 2 miles and stands out easily, even in this rugged region. You should have no trouble finding it 5 days after new Moon. I find no detail on its small floor.

Kant (19 miles) is a prominent crater about 1½ miles deep. There is a low central peak and the walls are terraced.

Taylor (25 × 21 miles) is an oval shaped crater. The terraced walls rise to almost 1½ miles above the floor. There is an easily seen central peak.

Delambre (32 miles) is a fine formation. It is almost 2 miles deep. The walls are terraced and there is a small central peak. See if you can make out the little 3 mile crater on its northern rim 6 days after new Moon.

Theon Jr. and Theon Sr. are as close to identical formations as you can get. They are both 11 miles across and just under 2 miles deep. I see no detail in these attractive little craters.

D. Spain, *The Six-Inch Lunar Atlas: A Pocket Field Guide*,
DOI 10.1007/978-0-387-87610-8_19, 

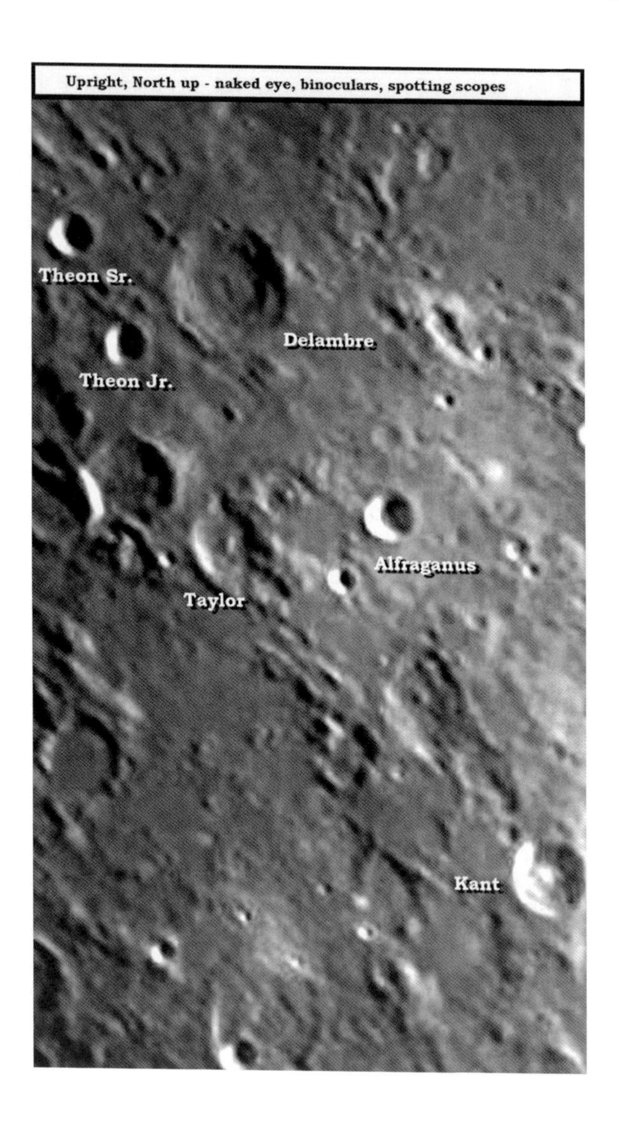
Upright, North up - naked eye, binoculars, spotting scopes
Theon Sr.
Delambre
Theon Jr.
Alfraganus
Taylor
Kant

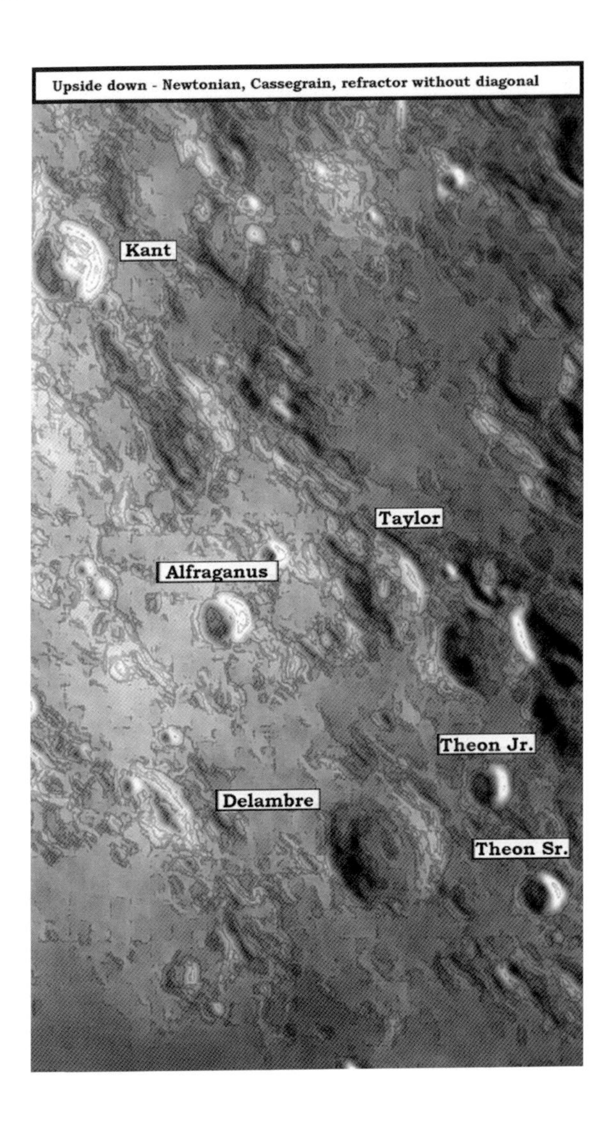
Upside down - Newtonian, Cassegrain, refractor without diagonal
Kant
Taylor
Alfraganus
Theon Jr.
Delambre
Theon Sr.

Upright, left/right reversed - refractors/catadioptrics with diagonal
Theon Sr.
Delambre
Theon Jr.
Alfraganus
Taylor
Kant

## CHART 18

# MUTUS

Mutus (47 miles) is a pretty large formation. Its walls reach up to over 2 miles from its large, flat floor. There are three craters within the walls. The largest is partially on the northeastern rim and is 14 miles across. On the southwest floor is a well formed 10 mile crater and on the eastern floor is a 9 mile crater. There are lots of detail visible in 4 inches of aperture. Look for it 5 days after new Moon.

Manzius (59 miles) has a rim about 2 miles above its large flat floor. There is an 11 mile crater on the southern rim. Look for long shadows when the sun is rising on the great formation.

Both of these craters are in the rugged and heavily cratered Southern Highlands. They are large enough and distinct enough that the careful observer will find they are not hard to identify, especially under a low to moderate sun.

D. Spain, *The Six-Inch Lunar Atlas: A Pocket Field Guide*,
DOI 10.1007/978-0-387-87610-8_20, 

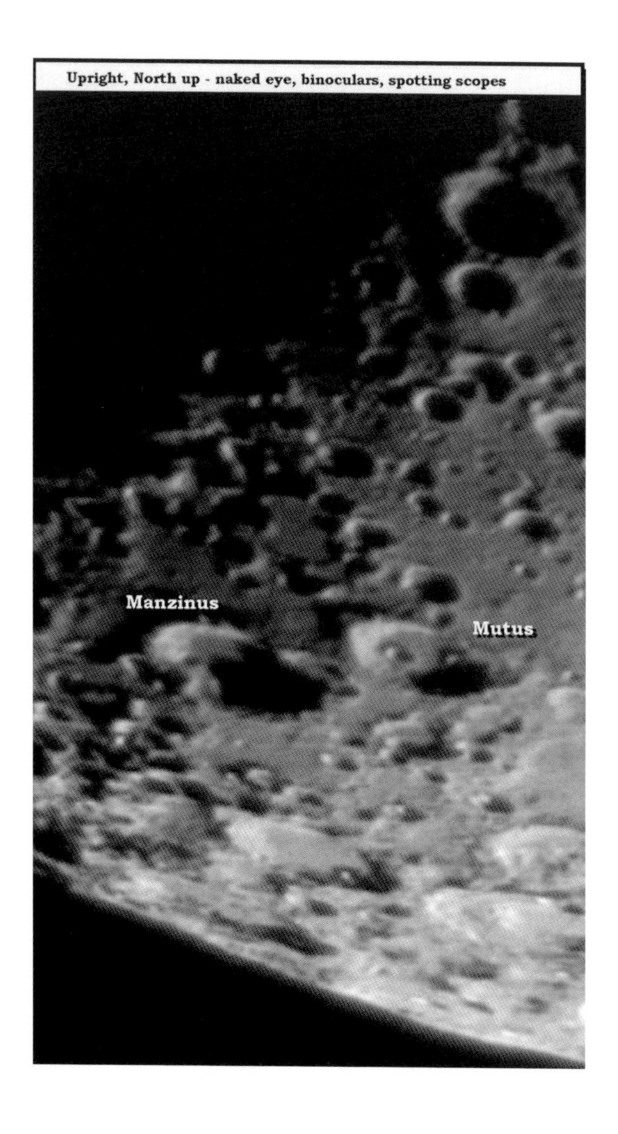
Upright, North up - naked eye, binoculars, spotting scopes
Manzinus
Mutus

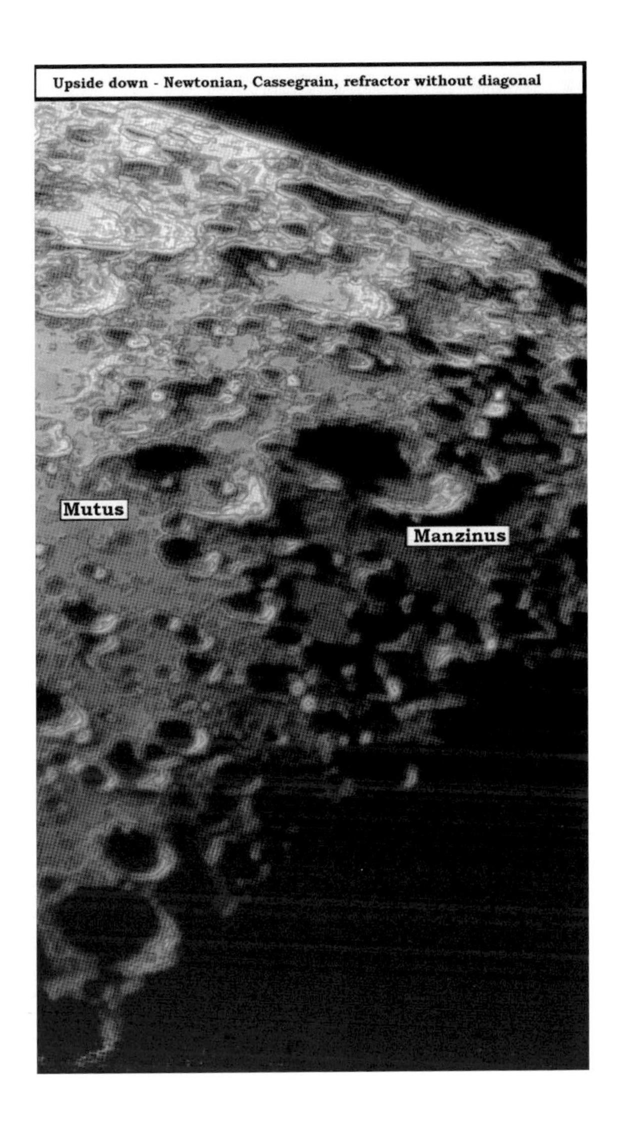
Upside down - Newtonian, Cassegrain, refractor without diagonal
Mutus
Manzinus

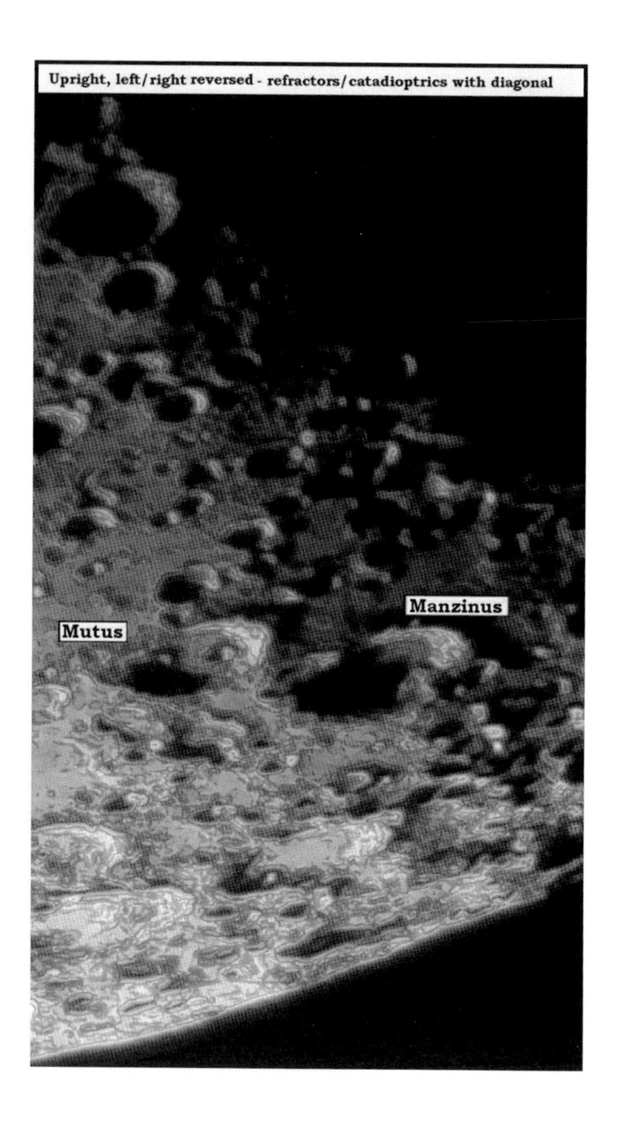
Upright, left/right reversed - refractors/catadioptrics with diagonal
Manzinus
Mutus

## CHART 19

# JULIUS CAESAR

Julius Caesar (55 miles) is a formation that is more or less rectangular in shape. Its eastern walls are battered and much lower than the western side. I estimate the western rim to be about a mile about the floor. A good part of the southeastern wall is non-existent. Notice that the large flat floor is much darker in the north. The best time to observe this region is 6 days after new Moon.

Sosigenes (11 miles) is a small, but prominent crater just east of Julius Caesar. It is 1 mile deep and has a featureless floor.

The area on the photo shows part of the shoreline of the Sea of Tranquility to the east and highlands to west. Under a rising sun you will see many hills and peaks on the higher elevation and the area abounds with rills and ridges.

D. Spain, *The Six-Inch Lunar Atlas: A Pocket Field Guide*,
DOI 10.1007/978-0-387-87610-8_21, 

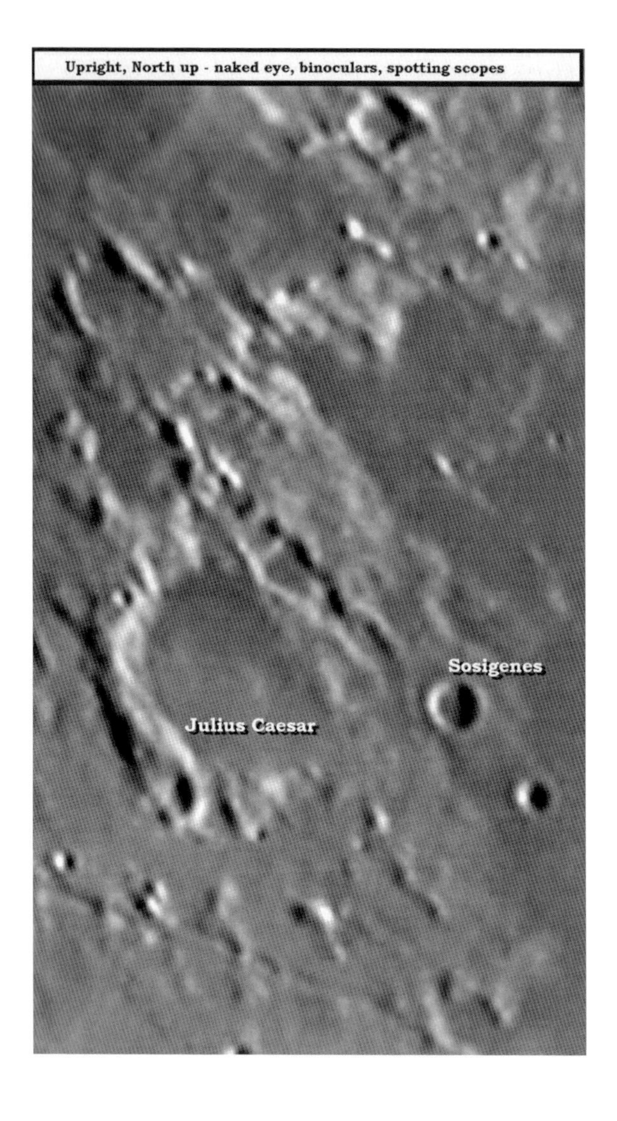
Upright, North up - naked eye, binoculars, spotting scopes
Sosigenes
Julius Caesar

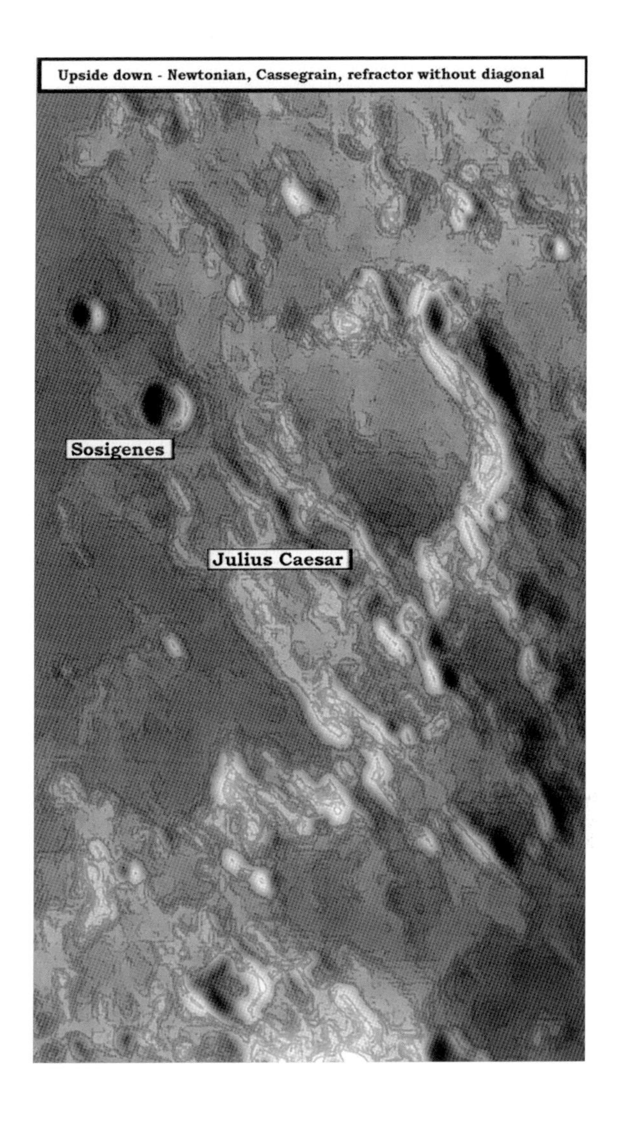
Upside down - Newtonian, Cassegrain, refractor without diagonal
Sosigenes
Julius Caesar

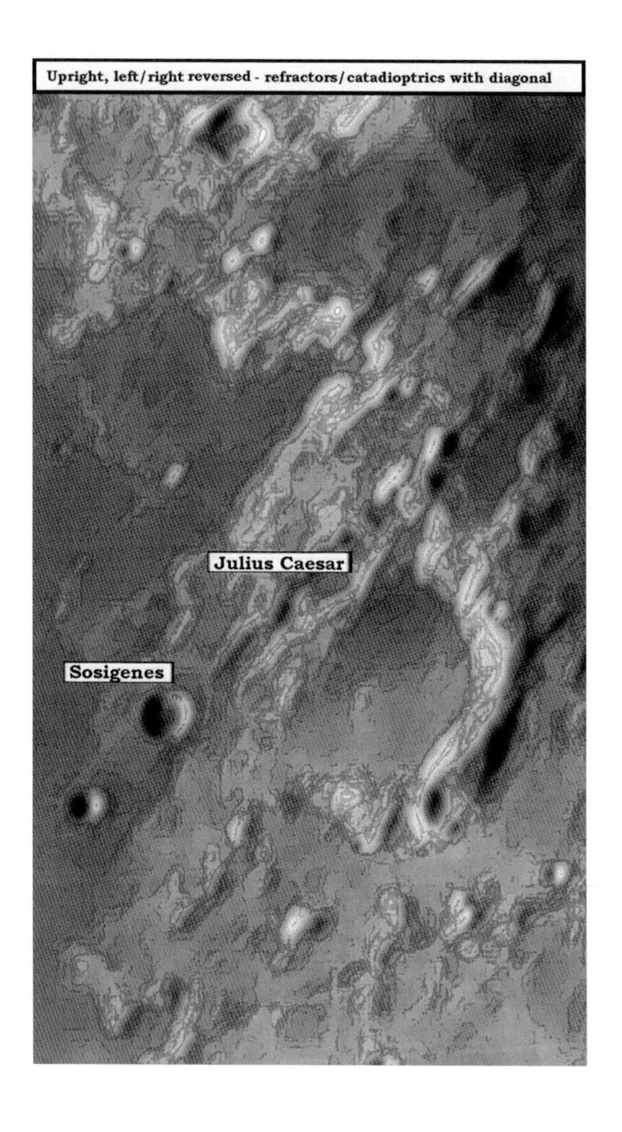
Upright, left/right reversed - refractors/catadioptrics with diagonal
Julius Caesar
Sosigenes

## CHART 20

# ALPINE VALLEY

The Alpine Valley (Vallis Alpes) is a great fault that splits the Alps Mountains. It is about 80 miles long and up to 7 miles wide. It is plainly visible in my 2.4 inch refractor at first quarter. With my 6 inch refractor I can sometimes make out parts of a rill that runs down most of the center of the valley. Be sure that you show this exceptional fault to the guests at any public viewing. It is sure to become one of your favorite areas on our sister world.

Protagoras (13 miles) is a sharp crater north of the valley. It is a little over a mile deep.

Archytas (19 miles) is another prominent crater. It is over a mile deep with terraced walls and an off center peak. The floor is flat and rough.

D. Spain, *The Six-Inch Lunar Atlas: A Pocket Field Guide*,
DOI 10.1007/978-0-387-87610-8_22, 

Upright, North up - naked eye, binoculars, spotting scopes
Archytas
Protagoras
Alpine Valley

Upside down - Newtonian, Cassegrain, refractor without diagonal
Alpine Valley
Protagoras
Archytas

Upright, left/right reversed - refractors/catadioptrics with diagonal
Archytas
Protagoras
Alpine Valley

## CHART 21

# ARISTILLUS

Aristillus (33 miles) is over 2 miles deep. This is a fine crater to view at first quarter Moon. On the flat floor is a complex mountain mass that rise to a little over ½ mile. The walls are terraced and take note of the lines of ridges that radiate outward from the base of this formation.

Autolycus (24 miles) is just 24 miles south of Aristillus and these formations make a pretty pair. It is 2 miles deep and the floor is rough, but I see no sign of hills or peaks on the floor. The walls are terraced and on the northeast rim is a small crater, 2 miles in diameter.

Theaetetus (15 miles) is a little crater west of the Caucasus Mountains. It is 1½ miles from rim top to the flat floor.

Cassini (35 miles) is a shallow crater about 20 miles northwest of Theaetetus. Its walls are less than a mile high and the floor is flat and flooded. There are two craters inside the walls of this formation. The largest is a ten mile crater that is about 1½ miles deep. The other is on the southwest floor and 5 miles across. Both are easily visible in my 2.4 inch refractor. These craters within Cassini make for a very interesting formation. Show it to your guests at public observations.

The Caucasus Mountains are to the east of the above formations. They are first visible a day before first quarter. The length of the range is about 300 miles and some of the higher peaks climb to a little over 2 miles above the surface.

D. Spain, *The Six-Inch Lunar Atlas: A Pocket Field Guide*,
DOI 10.1007/978-0-387-87610-8_23, © Springer Science+Business Media, LLC 2009

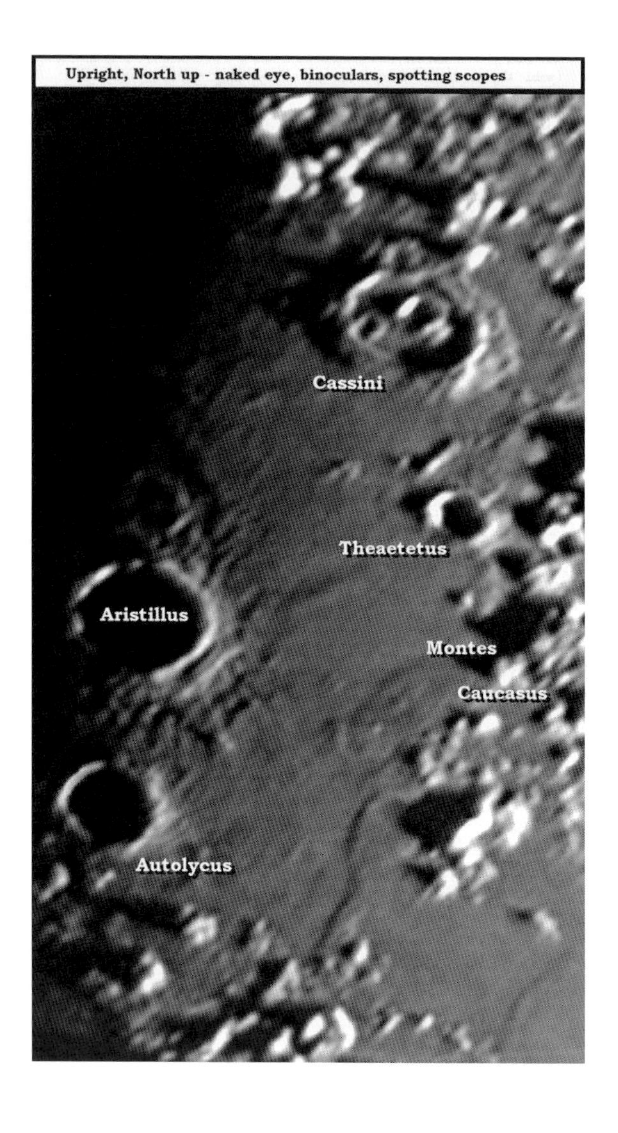
Upright, North up - naked eye, binoculars, spotting scopes
Cassini
Theaetetus
Aristillus
Montes
Caucasus
Autolycus

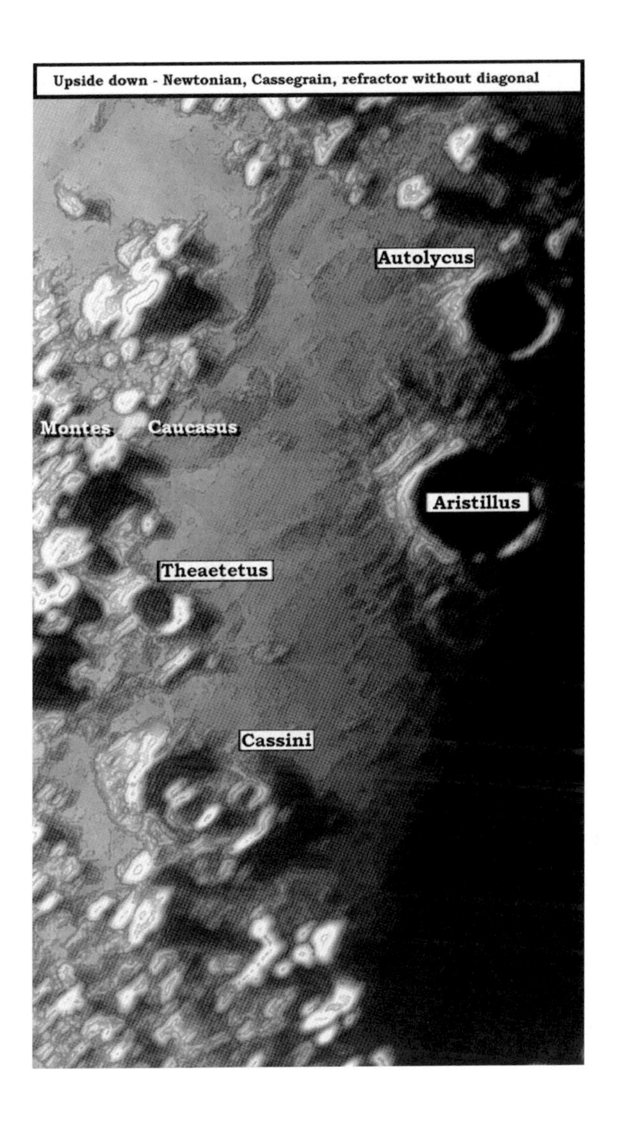
Upside down - Newtonian, Cassegrain, refractor without diagonal
Autolycus
Montes Caucasus
Aristillus
Theaetetus
Cassini

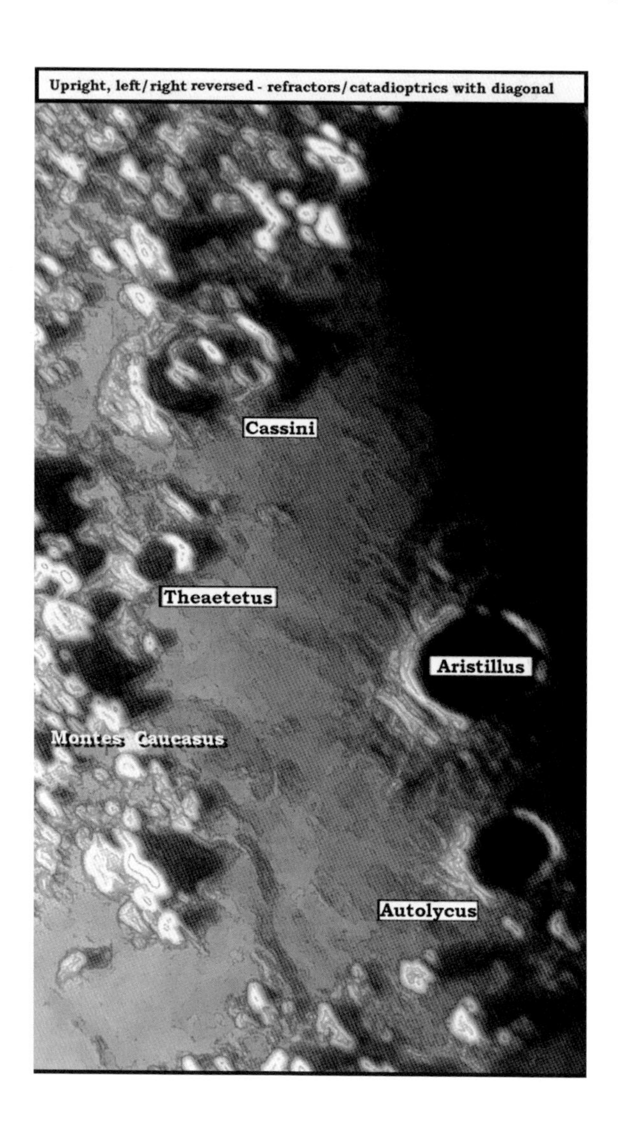
Upright, left/right reversed - refractors/catadioptrics with diagonal
Cassini
Theaetetus
Aristillus
Montes Caucasus
Autolycus

## CHART 22

# HYGINUS

Hyginus (8 miles) is a small crater about a ½ mile deep with a flat floor. While not particularly important by itself it is near the center of a remarkable rill named Rima Hyginus. One branch runs northwest from Hyginus and another branch runs to the east. The total length is about 132 miles. I have seen this attractive rill with my 2.4 inch refractor at 120× (6 days after new Moon). There is another rill system, visible on the photo, to the east of Rima Hyginus. This rill is not named on the photo, but it is Rima Ariadaeus. It is nearly the same length as the Hyginus rill.

Triesnecker (16 miles) is a sharp rimmed crater with a central peak. It is 1½ mile deep and the walls are somewhat terraced. There is another system of rills to the east of Triesnecker, running mostly north and south. The deeper rills are visible in my 2.4 inch refractor, but to fully explore the system 6 inches or more of aperture is needed.

Agrippa (28 miles) is a prominent crater, almost 2 miles deep with a central peak. The walls are terraced and there is a little 4 mile crater on the north rim.

Godin (21 miles) is another prominent crater just under 2 miles deep. There is an off center peak on the rough floor.

Be sure to show guests at a public gathering the many rills in this area when it is illuminated by grazing to moderate sunlight.

D. Spain, *The Six-Inch Lunar Atlas: A Pocket Field Guide*,
DOI 10.1007/978-0-387-87610-8_24, 

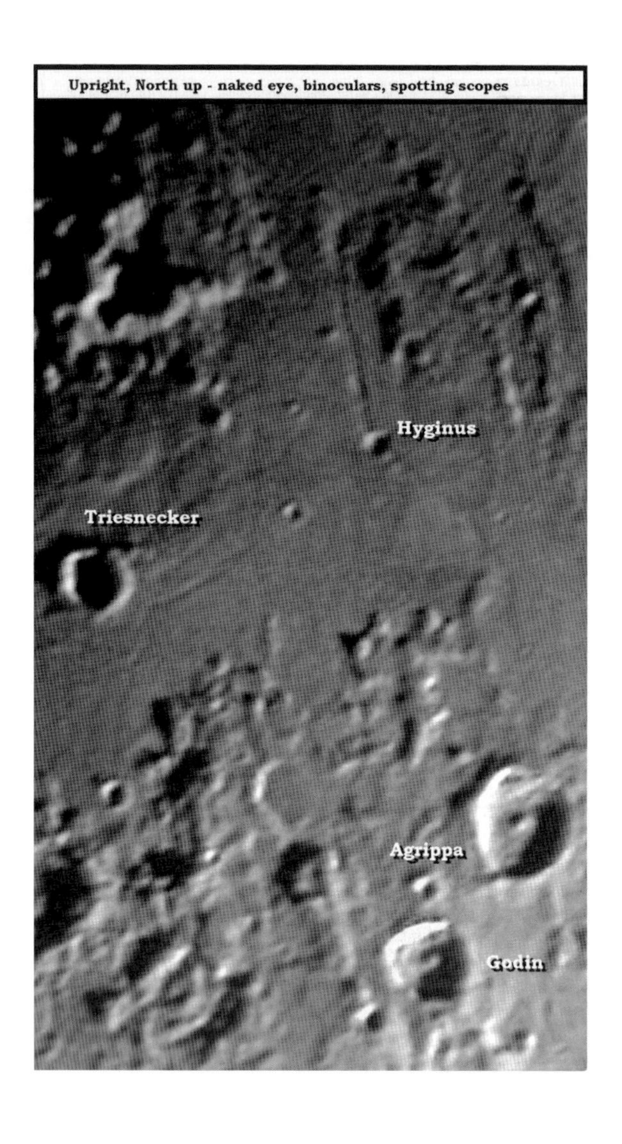
Upright, North up - naked eye, binoculars, spotting scopes
Hyginus
Triesnecker
Agrippa
Godin

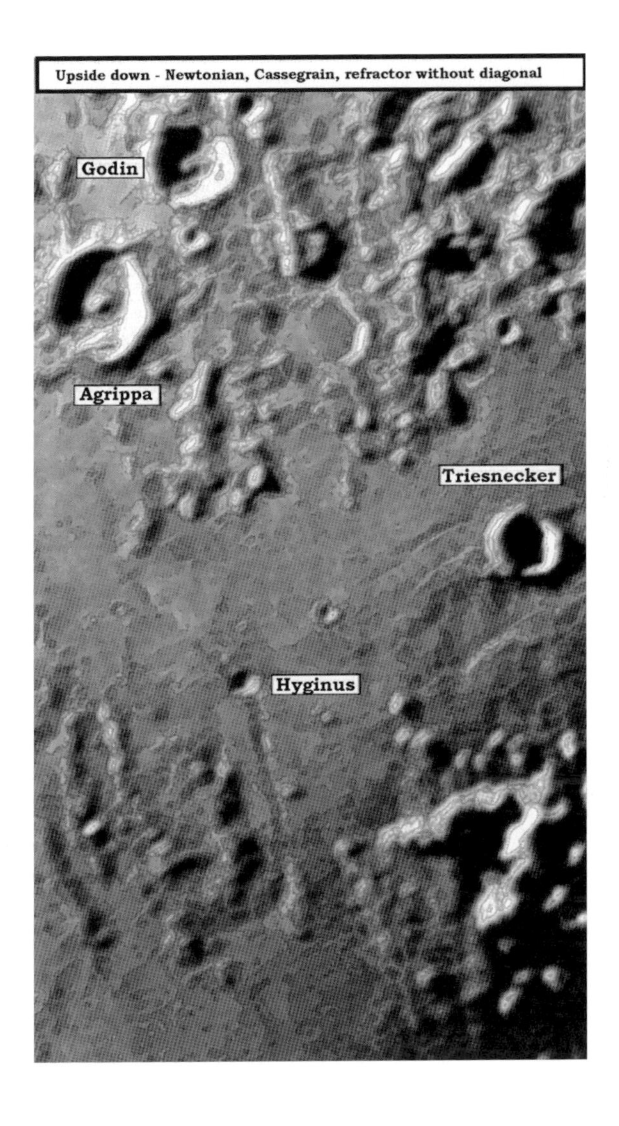
Upside down - Newtonian, Cassegrain, refractor without diagonal
Godin
Agrippa
Triesnecker
Hyginus

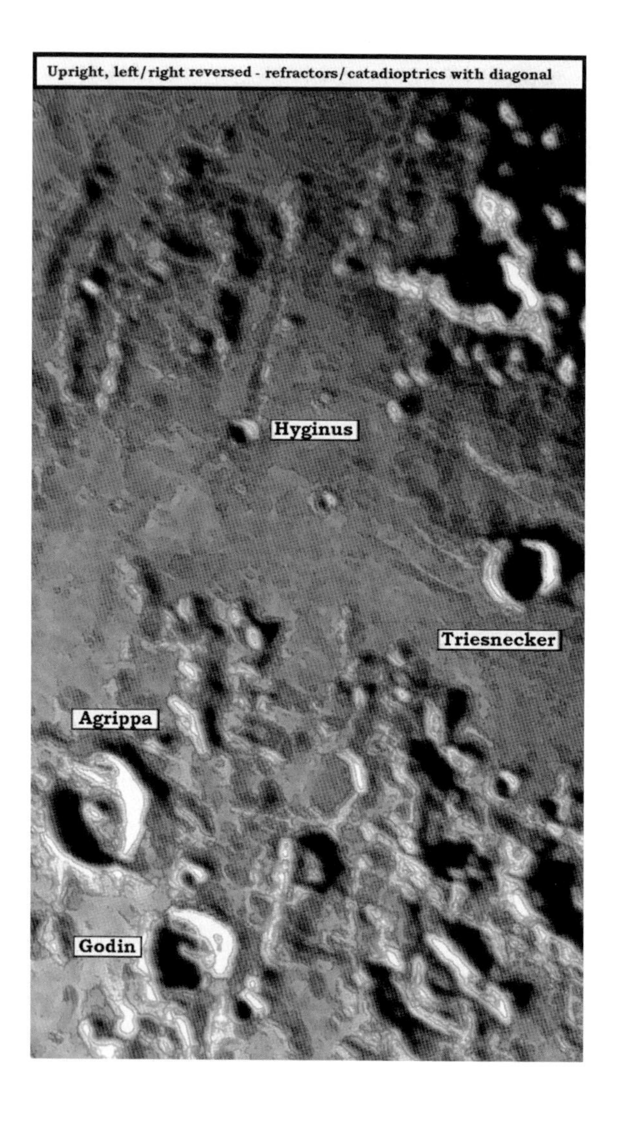
Upright, left/right reversed - refractors/catadioptrics with diagonal
Hyginus
Triesnecker
Agrippa
Godin

## CHART 23

# HIPPARCHUS

Hipparchus (90 miles) is a battered and worn down formation. The highest walls are on the southwest rim and may reach 1½ miles above the floor. The northwestern rim is very low and missing in some places. The very large flat floor is full of detail as the sun illuminates this region at first quarter.

Horrocks (18 miles) is a very prominent crater on the northeastern floor of Hipparchus. It is a fine crater in its own right. The walls are terraced and rise almost 2 miles above a small central peak.

Halley (22 miles) is about 1½ miles deep. There is a tiny central peak, really just a hill that is visible as the shadow of the eastern wall moves off of it at sunrise.

Hind (18 miles) is a fine crater just east of Halley. Not quite 2 miles deep, I can find no evidence of a central peak on the rough floor.

Albategnius (82 miles) has well formed walls that rise almost 2 miles above the vast flat floor. There is a prominent 12 mile crater on the north floor. Notice the small 4 mile crater on the outside of this crater's southwestern rim. The floor has a beautiful off center peak just under a mile high. It casts a long shadow onto the crater Klein at sunrise.

Klein (27 miles) has destroyed Albategnius' southwestern wall. It is less than a mile deep, has a flat floor and a small central peak.

This is a fascinating area, well worth your time to explore at first quarter.

D. Spain, *The Six-Inch Lunar Atlas: A Pocket Field Guide*,
DOI 10.1007/978-0-387-87610-8_25, 

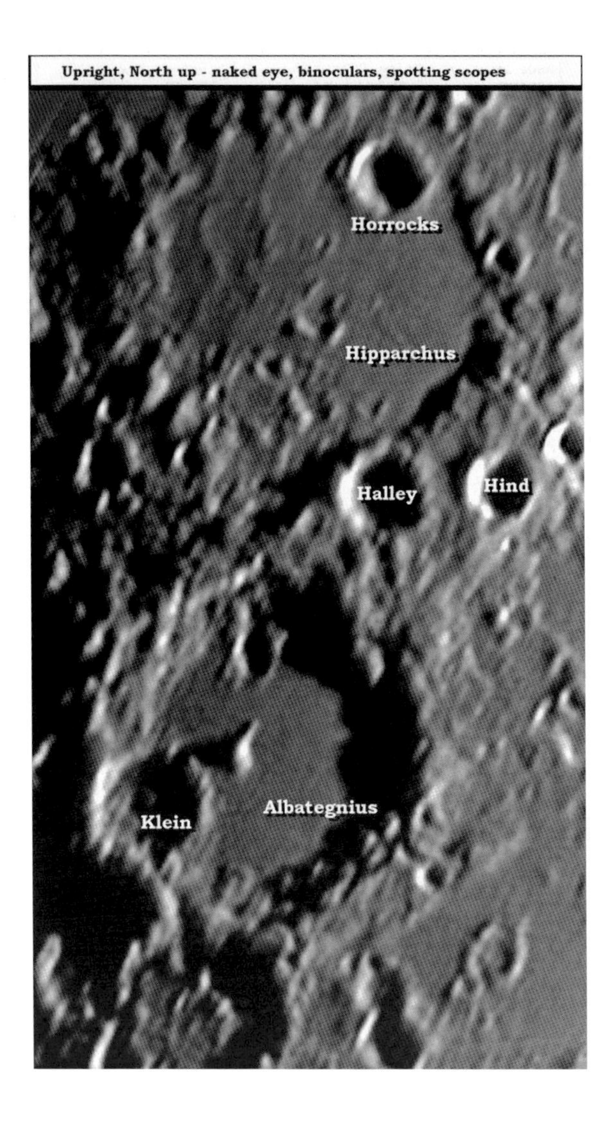
Upright, North up - naked eye, binoculars, spotting scopes
Horrocks
Hipparchus
Halley
Hind
Klein
Albategnius

Upside down - Newtonian, Cassegrain, refractor without diagonal
Klein
Albategnius
Halley
Hind
Hipparchus
Horrocks

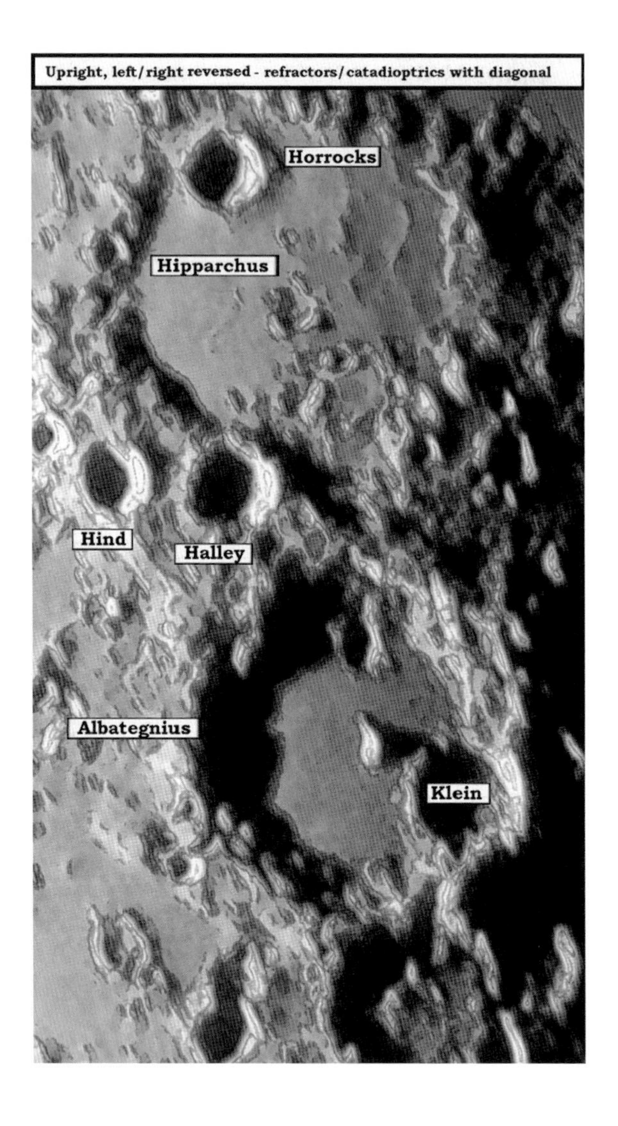

Upright, left/right reversed - refractors/catadioptrics with diagonal
Horrocks
Hipparchus
Hind
Halley
Albategnius
Klein

## CHART 24

# HERACLITUS

Heraclitus (55 miles) is fairly well worn down with some parts of its rim reaching upwards to over 1½ miles. Its north rim is completely destroyed by Licetus and its southwestern rim is similar damaged by a 31 mile crater. Instead of a central peak, there is a central crest running northeast from the rim of an intruding southern crater almost to the rim of Licetus. I would estimate its height at less than ½ mile. The floor is rough and my 6 inch refractor reveals several small craters on the eastern side of the central crest. Find this one 6 days after new Moon.

Licetus (45 miles) is a great circular crater, over 2 miles deep. There are two sizable craters that have destroyed part of its southern rim. The floor contains many small craters and a small central hill.

Cuvier (45 miles) is another circular crater, also over 2 miles deep. It has a large flat floor that has little of interest.

Lilius (37 miles) is a prominent crater not quite 2 miles deep. The walls are somewhat terraced and there is a large central mountain that dominates the floor and climbs up to about ¾ mile.

These craters are in the heavily cratered Southern Highlands, but with careful searching you should have no trouble findings these fine formations. The central crest in Heraclitus will tell you that you are in the right area.

D. Spain, *The Six-Inch Lunar Atlas: A Pocket Field Guide*,
DOI 10.1007/978-0-387-87610-8_26, © Springer Science+Business Media, LLC 2009

Upright, North up - naked eye, binoculars, spotting scopes
Licetus
Heraclitus
Cuvier
Lilius

Upside down - Newtonian, Cassegrain, refractor without diagonal
Lilius
Cuvier
Heraclitus
Licetus

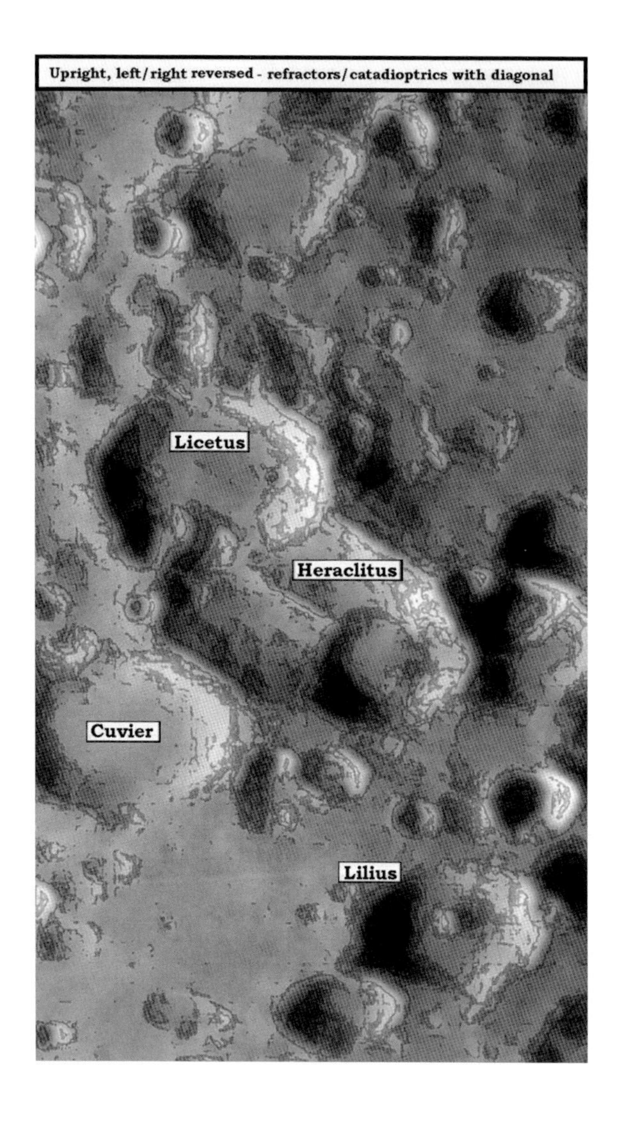
Upright, left/right reversed - refractors/catadioptrics with diagonal
Licetus
Heraclitus
Cuvier
Lilius

## CHART 25

# PLATO

Plato (61 miles) is a great flooded crater. Its walls are not particularly high, with just a few parts of the rim reaching up 1½ miles. The floor is dark and the lava that has flooded it has dissolved any traces of a central peak or hills. There are many craterlets and pits on the floor that are best seen 1 day after first quarter under a rising sun. In my 6 inch refractor I can pick out six of them under good seeing conditions. Look for long shadows on the floor as the sun rises over this magnificent formation. It is a must show at any public observation.

Piazza Smyth (13 miles) is a small, but very prominent crater. It is about 1½ miles deep with a featureless floor.

Montes Alps, the Alps Mountains surround Plato. This range of mountains is scattered on the island-like highlands that separate Mare Imbrium, the Sea of Rains from Mare Frigoris, the Sea of Cold. The higher peaks reach to about 1½ miles.

Montes Teneriffe, the Teneriffe Mountains lay south of Plato out in the Sea of Rains. They are a loosely scattered collection of peaks, the higher ones 1½ miles above the plain.

Mons Pico or just Pico is an isolated peak to the east of the Teneriffes. It is about 1½ miles and when catching the rays of a lunar sunrise casts a great long shadow.

This is a wonderful area of mixed features. There are many wrinkle ridges that are easily view with a 2.4 inch telescope.

D. Spain, *The Six-Inch Lunar Atlas: A Pocket Field Guide*,
DOI 10.1007/978-0-387-87610-8_27, © Springer Science+Business Media, LLC 2009

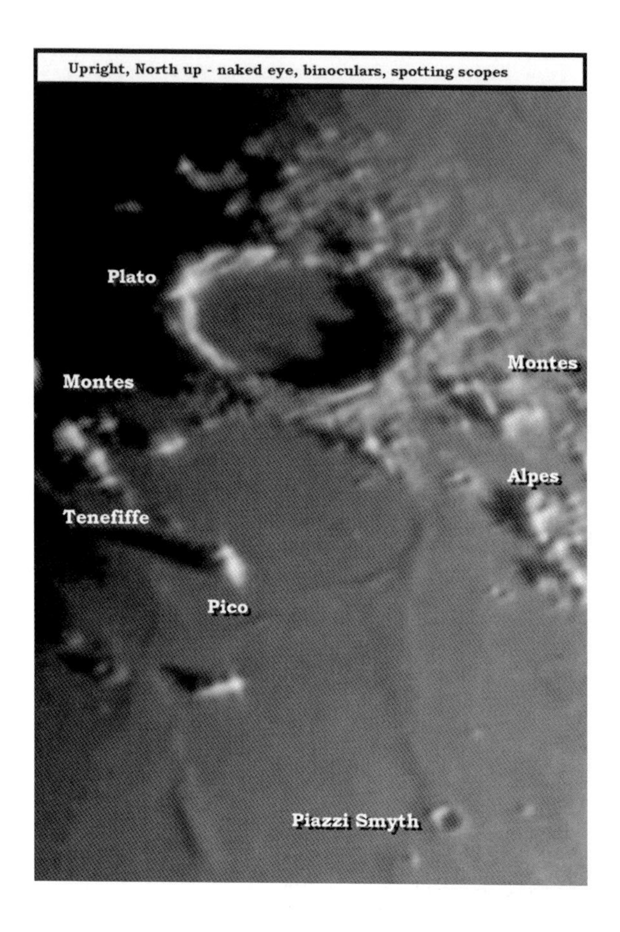
Upright, North up - naked eye, binoculars, spotting scopes
Plato
Montes
Montes
Alpes
Tenefiffe
Pico
Piazzi Smyth

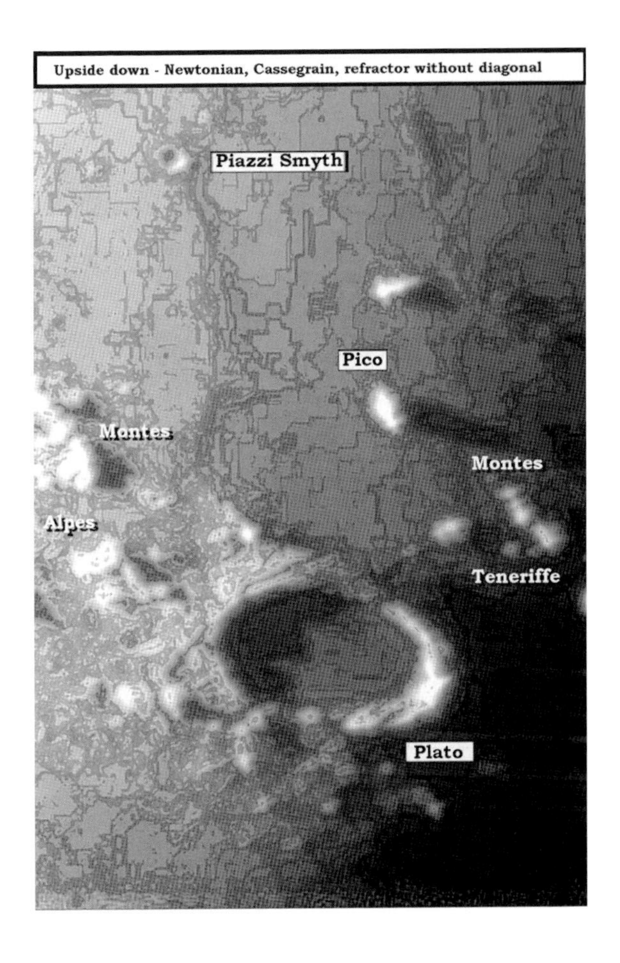
Upside down - Newtonian, Cassegrain, refractor without diagonal
Piazzi Smyth
Pico
Montes
Montes
Alpes
Teneriffe
Plato

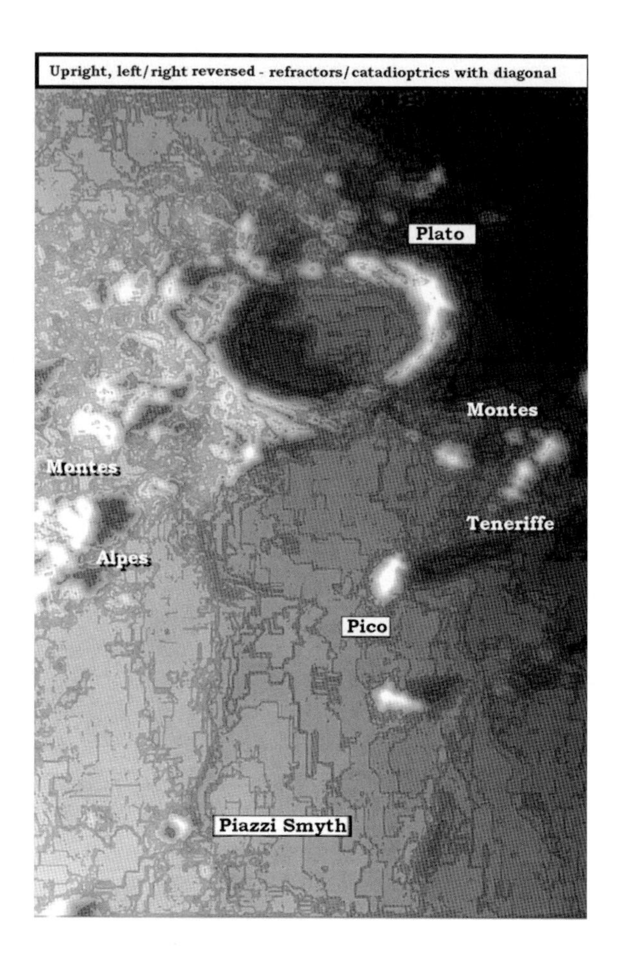
Upright, left/right reversed - refractors/catadioptrics with diagonal
Plato
Montes
Montes
Teneriffe
Alpes
Pico
Piazzi Smyth

## CHART 26

# ARCHIMEDES

Archimedes (50 miles) is a great classic crater. Even 2.4 inches of aperture will give pleasurable views. This fine circular structure has a flat floor that has been flooded with lava. The walls are terraced and rise to a little over a mile above the floor. This beautiful formation is a nice showpiece for any public observation at first quarter. At a magnification of 240 it is very impressive in my 6 inch refractor. No craters or hills are visible on the floor. When the sun is just rising over the eastern ramparts you will see several long shadows cast over the floor.

Bancroft (8 miles) is a sharp crater to the southwest of Archimedes. It is just under a 1½ mile deep and I see no detail on its small floor.

Beer (6 miles) and Feuillee (6 miles) are two almost identical craters. Beer is just a little shallower with both being about 1 mile deep. They make a very attractive pair. Because of there similarity maybe they should have been named Castor and Pollux, the Gemini twins.

Montes Archimedes, the Archimedes Mountains is a pretty little range of mountains. The peaks are not particularly high with none over 2 miles high. Nevertheless, under a low sun this little range is a grand sight.

D. Spain, *The Six-Inch Lunar Atlas: A Pocket Field Guide*,
DOI 10.1007/978-0-387-87610-8_28, 

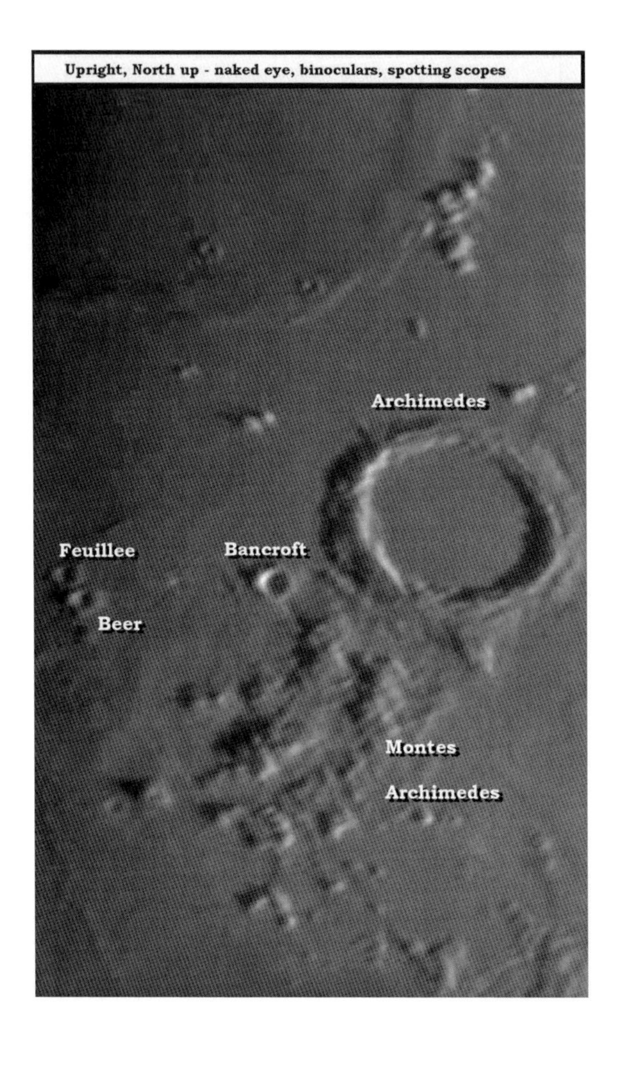
Upright, North up - naked eye, binoculars, spotting scopes
Archimedes
Feuillee
Bancroft
Beer
Montes
Archimedes

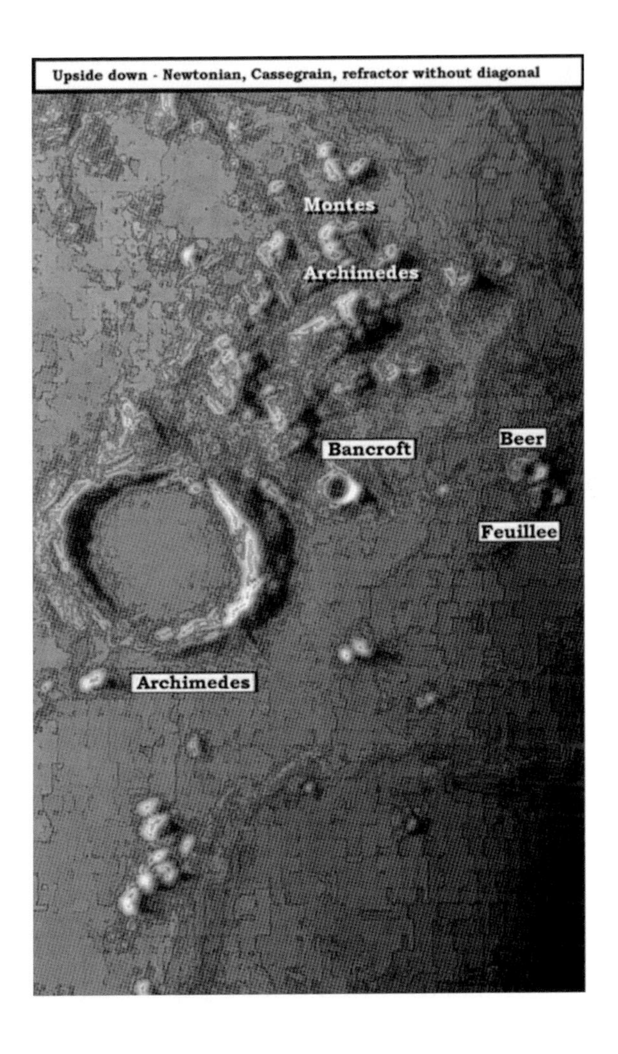
Upside down - Newtonian, Cassegrain, refractor without diagonal
Montes
Archimedes
Bancroft
Beer
Feuillee
Archimedes

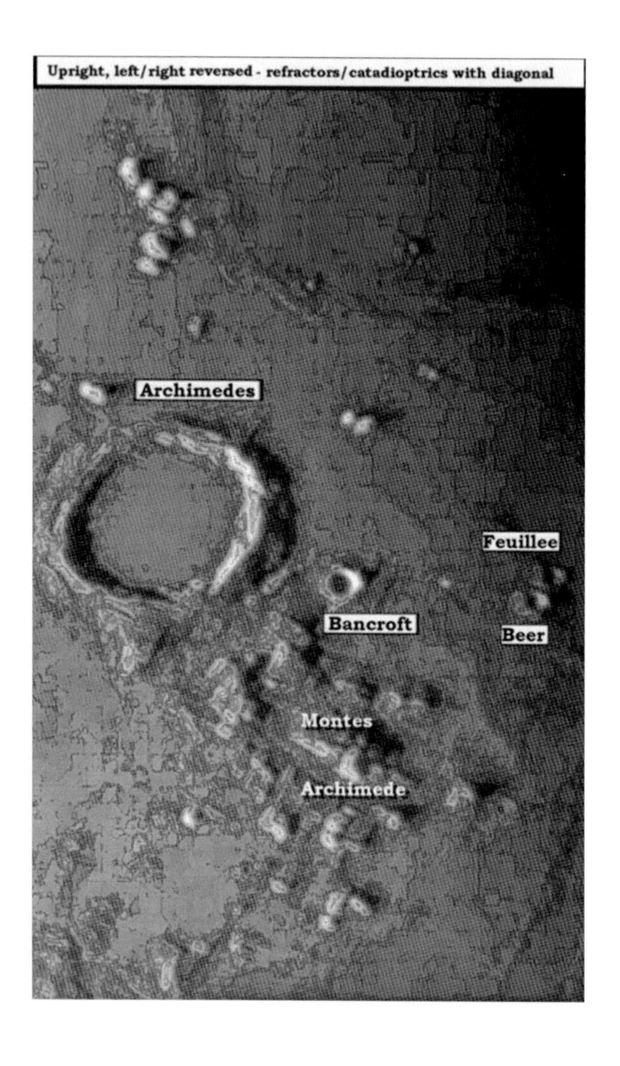
Upright, left/right reversed - refractors/catadioptrics with diagonal
Archimedes
Feuillee
Bancroft
Beer
Montes
Archimede

CHART 27

## APENNINE MOUNTAINS

Montes Apenninus, the Apennine Mountains are the largest and longest of the lunar mountains. The southern section is imaged on the photo. The peaks rise rapidly on the west from the Sea of Rains and slowly taper off to foothills toward the east. A must sight for you and your guest under a rising sun.

Mons Wolff, Wolff Mountain, Mons Ampere, Ampere Mountain and Mons Huygens, Huygens Mountain are three great peaks in the Apennines. Wolff juts up just over 2 miles above the plain below. Ampere is the smallest of the three peaks, rising to about 1¾ miles. Huygens is the great mountain mass north of Ampere. It towers over 3 miles above the Sea of Raines.

Eratosthenes (35 miles) is a lovely crater at the southern end of the Apennines Mountains. Its rim is 2 miles above the floor and the walls are terraced. It has a great central mountain massif with three distinct peaks. Observe and show it off to your guests 1 day after first quarter.

Wallace (11 miles) is a square shaped crater. Its eastern wall is almost entirely gone and the floor is completely flooded. The walls are highest in the south, but I doubt if they are more than a few hundred feet high.

This is another truly fascinating region that you will fall in love with and you will come back time and time again. Beautiful with 2.4 inches of aperture, it is absolutely stunning in my 6 inch refractor.

D. Spain, *The Six-Inch Lunar Atlas: A Pocket Field Guide*,
DOI 10.1007/978-0-387-87610-8_29, 

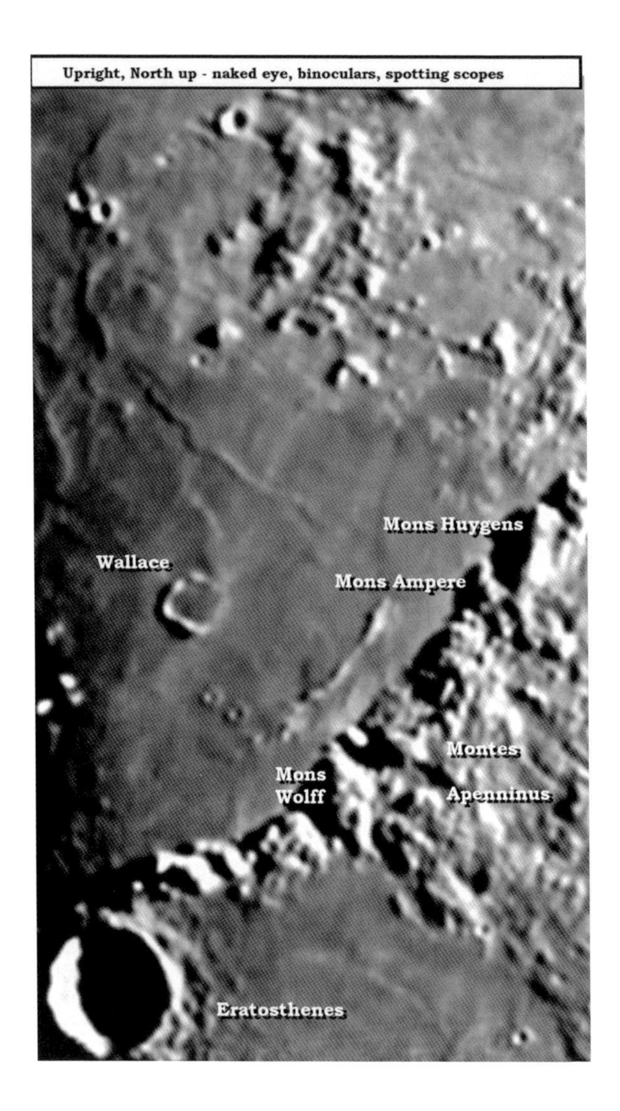
Upright, North up - naked eye, binoculars, spotting scopes
Mons Huygens
Wallace
Mons Ampere
Montes
Apenninus
Mons
Wolff
Eratosthenes

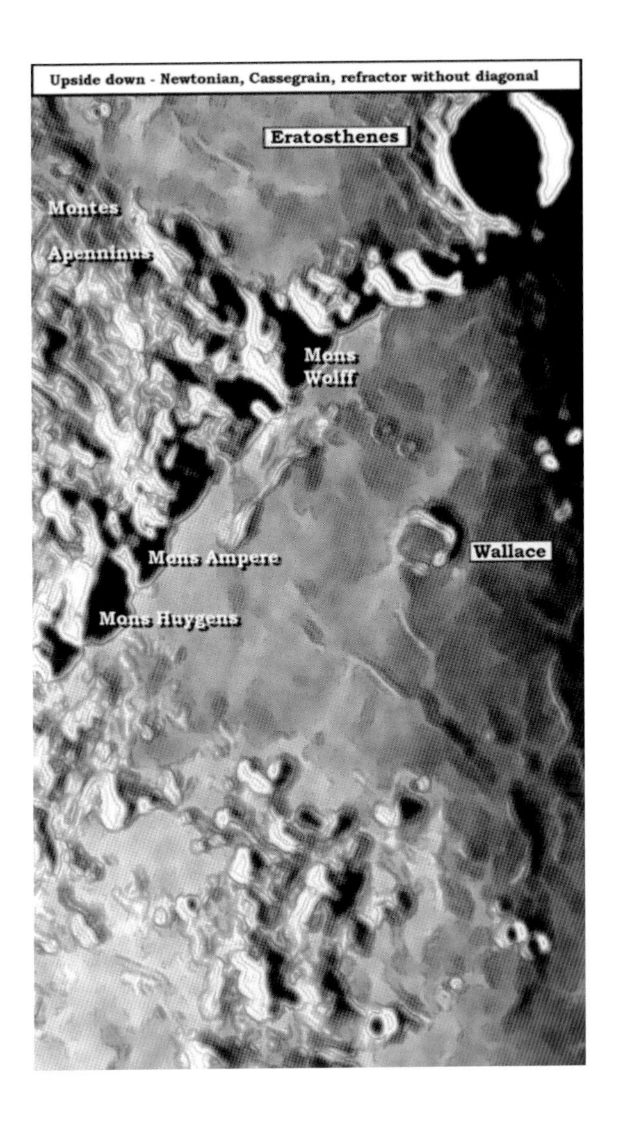
Upside down - Newtonian, Cassegrain, refractor without diagonal
Eratosthenes
Montes
Apenninus
Mons
Wolff
Wallace
Mons Ampere
Mons Huygens

Upright, left/right reversed - refractors/catadioptrics with diagonal
Mons Huygens
Mons Ampere
Wallace
Mons
Wolff
Montes
Apenninus
Eratosthenes

## CHART 28

# SCHROTER

Schroter (16 miles) is a partial ruined formation that is flooded and almost bisected. The east and west walls are low, with the eastern peak probably less than a thousand feet high.

Sommering (17 miles) is another ruined formation, flooded and split into just the eastern and western walls. Like Schroter, the eastern side is the larger and appears to be about ½ mile high.

The area north of these craters is a fascinating area. The topology of the region is low hills and small craters and dark color. There are several areas on the Moon that look like they are coated with a dark mantel, but this is one of the largest. Note the roughness here. Parts of it almost look like gigantic open pit coal mining. Of course there is no coal on the Moon, but if you are observing 1 day after first quarter look be sure to explore here. This dark area is easily visible in 7 × 35 binoculars.

D. Spain, *The Six-Inch Lunar Atlas: A Pocket Field Guide*,
DOI 10.1007/978-0-387-87610-8_30, 

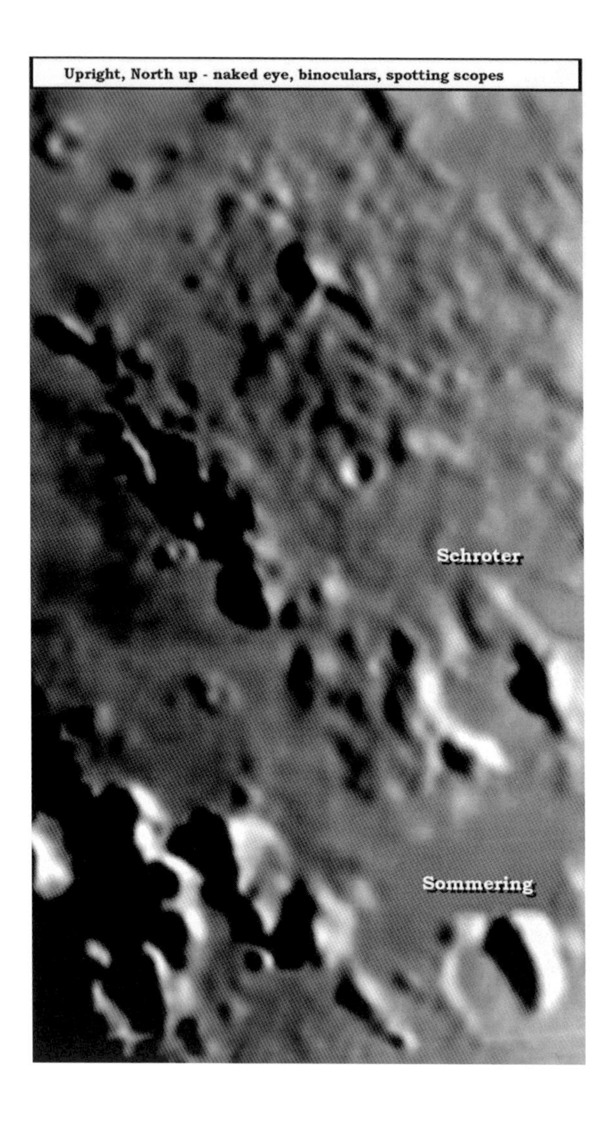
Upright, North up - naked eye, binoculars, spotting scopes
Schroter
Sommering

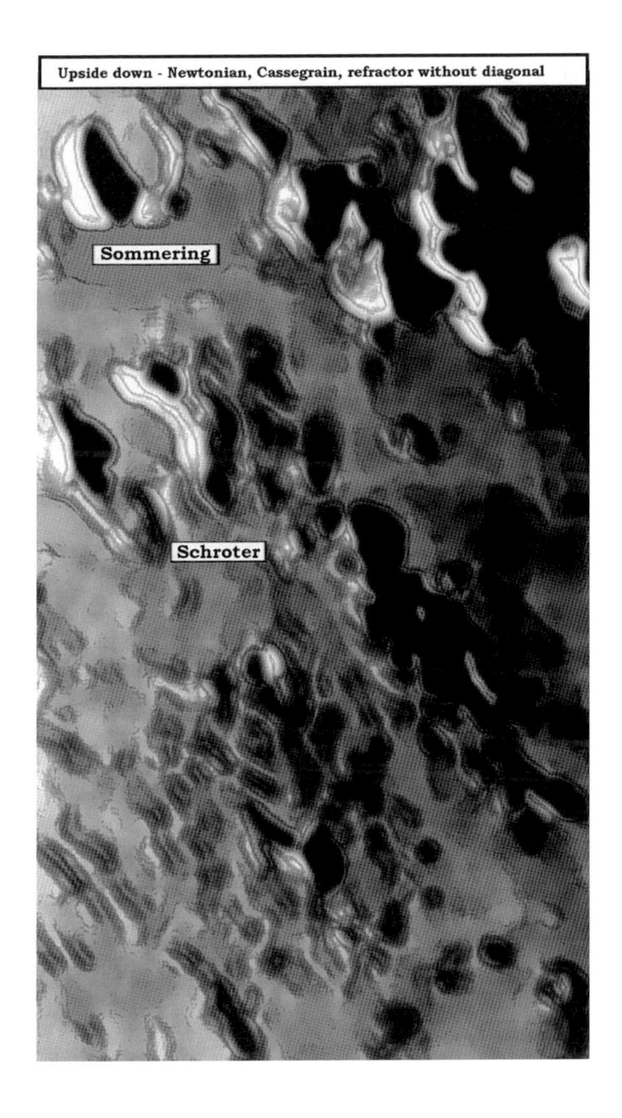
Upside down - Newtonian, Cassegrain, refractor without diagonal
Sommering
Schroter

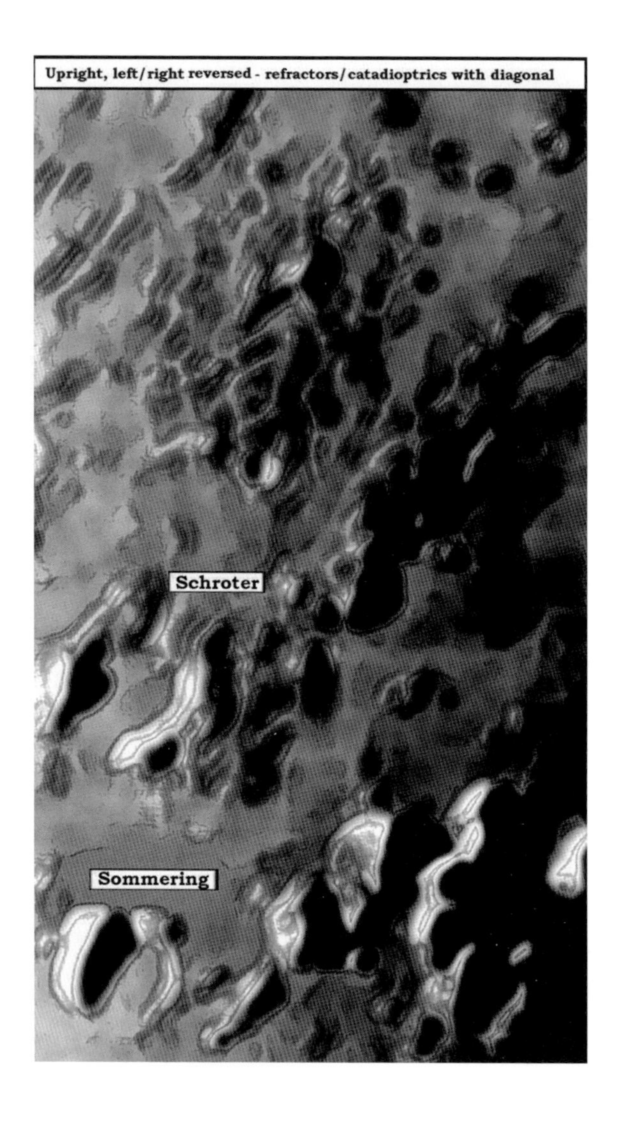
Upright, left/right reversed - refractors/catadioptrics with diagonal
Schroter
Sommering

## CHART 29

# PTOLEMAEUS

Ptolemaeus (93 miles) is a great circular formation a little south and just west of the center of the Moon. As the sun rises on this old giant at first quarter, long shadows are cast across the great expanse of the flat floor. Other than Ammonius, there are no sizable craters on the floor and it lacks any hills or crests. However, the floor is not without interest. Under low grazing illumination, you will notice several shallow, saucer shaped depressions. The largest of these depressions is directly north of Ammonius and has a diameter of about 10 miles. The great walls that surround the vast interior plain rise above the floor to more than 1½ miles in places. This old monster is a prize formation to show off to anyone at public observations. This formation will become one of your favorites, visible in any telescope and easily visible in 7 × 35 binoculars.

Ammonius (5 miles) is the prominent crater on the northeastern floor of Ptolemaeus. Though small in diameter, it is a little over a mile deep.

D. Spain, *The Six-Inch Lunar Atlas: A Pocket Field Guide*,
DOI 10.1007/978-0-387-87610-8_31, 

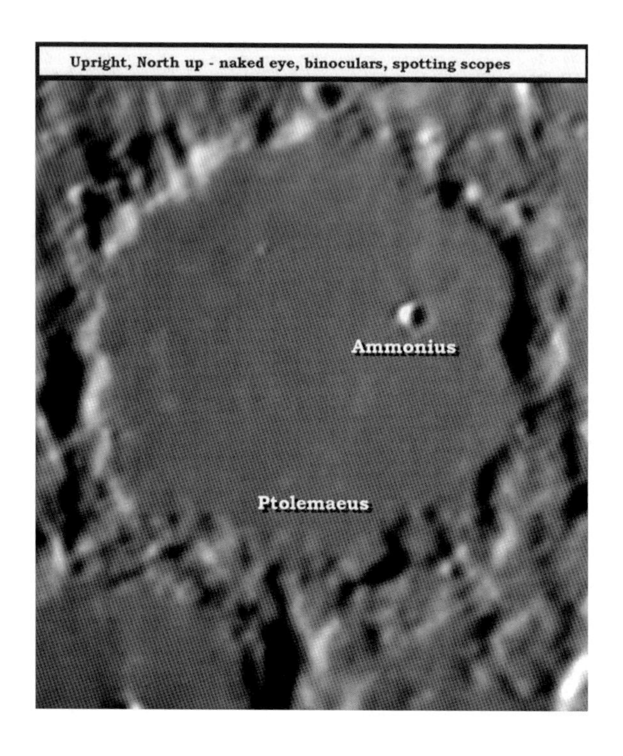
Upright, North up - naked eye, binoculars, spotting scopes
Ammonius
Ptolemaeus

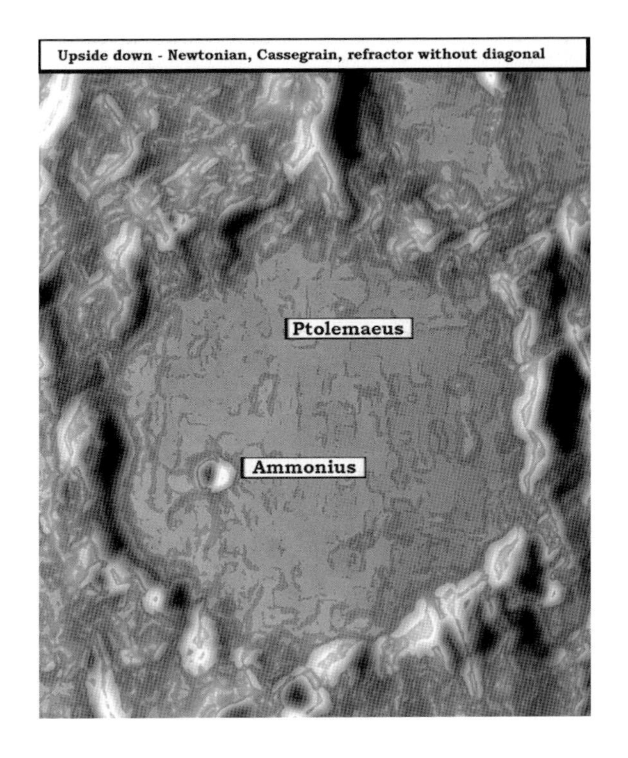
Upside down - Newtonian, Cassegrain, refractor without diagonal
Ptolemaeus
Ammonius

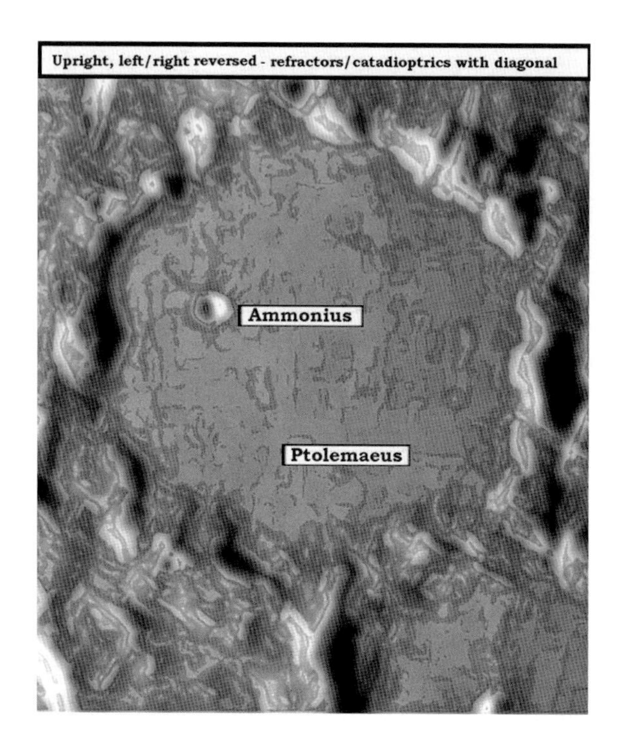
Upright, left/right reversed - refractors/catadioptrics with diagonal
Ammonius
Ptolemaeus

CHART 30

# ALPHONSUS

Alphonsus (71 miles) is directly south of Ptolemaeus and its northern wall abuts on Ptolemaeus' southern wall. Its walls are 1½ miles above the large flat floor. There is a small, but prominent central peak that is a little over a ½ mile high. The floor has some dark regions; the largest is on the western floor. The eastern floor has an extensive system of rills. This formation is a must show for yourself and your guests at first quarter.

Arzachel (59 miles) is the third of three great craters in a north-south alignment. Ptolemaeus and Alphonsus are its big brothers, but is no slouch itself. It is the deepest of the three with terraced walls that tower over 2 miles about its flat floor. There is a rill system on the eastern floor and an elongated central mountain mass that is over ½ mile high. The floor is rough with hills and small craters. The crater due east of the central mountain is 6 miles across and about a mile deep.

Alpetragius (24 miles) has an unusually massive central massif. This mountain just about occupies the entire floor of this crater that is over 2 miles deep.

These formations, along with Ptolemaeus make a very attractive line of craters that must be shown to anyone who looks through your eyepiece.

D. Spain, *The Six-Inch Lunar Atlas: A Pocket Field Guide*,
DOI 10.1007/978-0-387-87610-8_32, © Springer Science+Business Media, LLC 2009

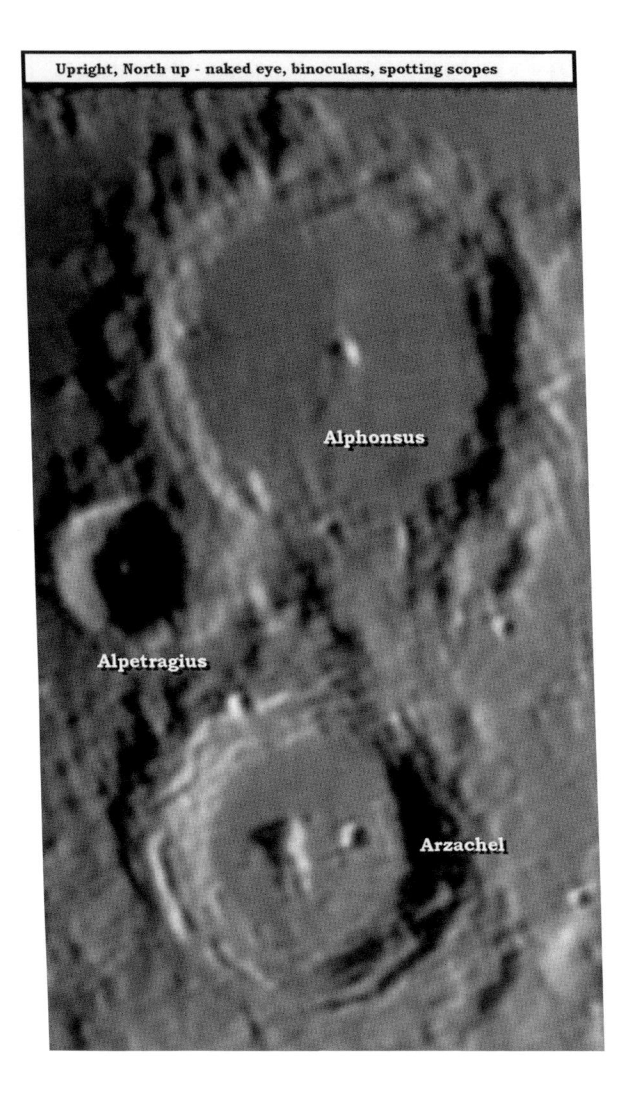
Upright, North up - naked eye, binoculars, spotting scopes
Alphonsus
Alpetragius
Arzachel

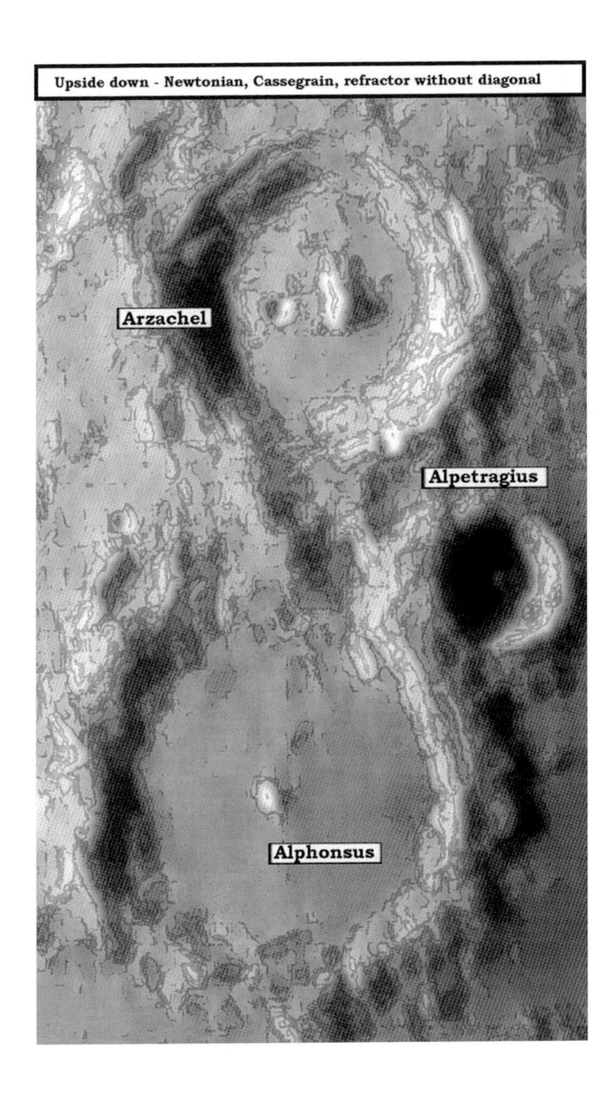
Upside down - Newtonian, Cassegrain, refractor without diagonal
Arzachel
Alpetragius
Alphonsus

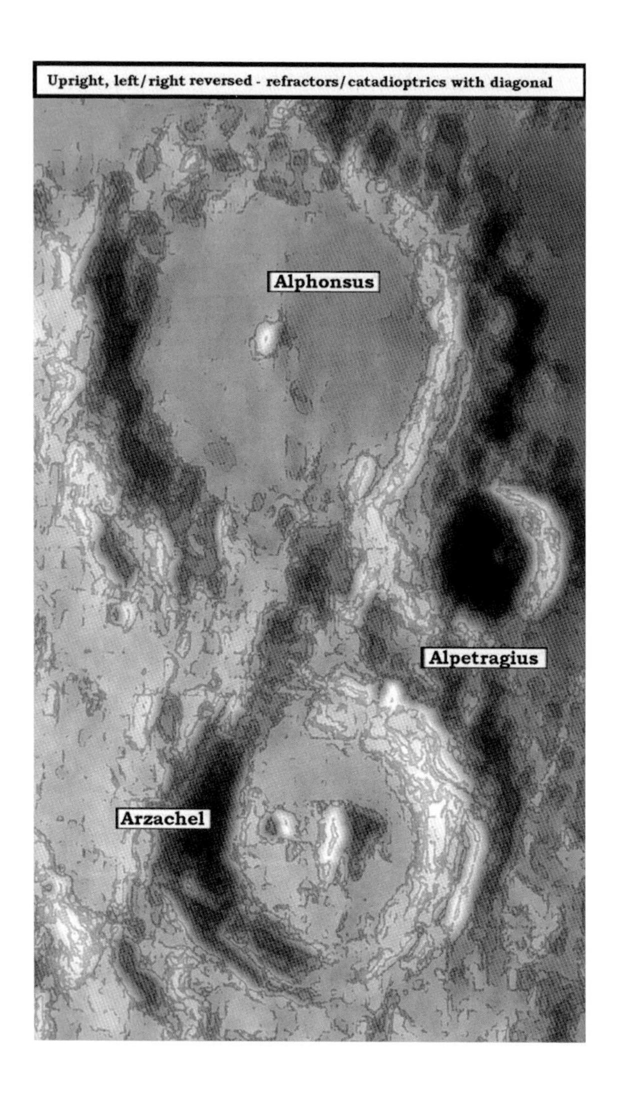
Upright, left/right reversed - refractors/catadioptrics with diagonal
Alphonsus
Alpetragius
Arzachel

## CHART 31

# WALTER

Walter (80 miles) is a large, flat floor crater with high terraced walls that exceed 2 miles. The floor is mainly smooth, except on the northeastern floor where there are hills and a grouping of five craters, each about the same size. Although this formation is located in the extensively cratered highlands you should have no problem locating it at first quarter.

Aliacensis (48 miles) is a fine crater over 2 miles deep with terraced walls. The flat floor has a small, off center peak.

Werner (42 miles) is a circular crater, similar to Aliacensis, but a little smaller. Its terraced walls rise 2 miles above the floor. There is a small peak that is off center and the rough floor has small hills.

These three craters make an attractive trio. An excellent target for smaller scopes, there is a wealth of detail in 4 inch and larger instruments.

D. Spain, *The Six-Inch Lunar Atlas: A Pocket Field Guide*,
DOI 10.1007/978-0-387-87610-8_33, © Springer Science+Business Media, LLC 2009

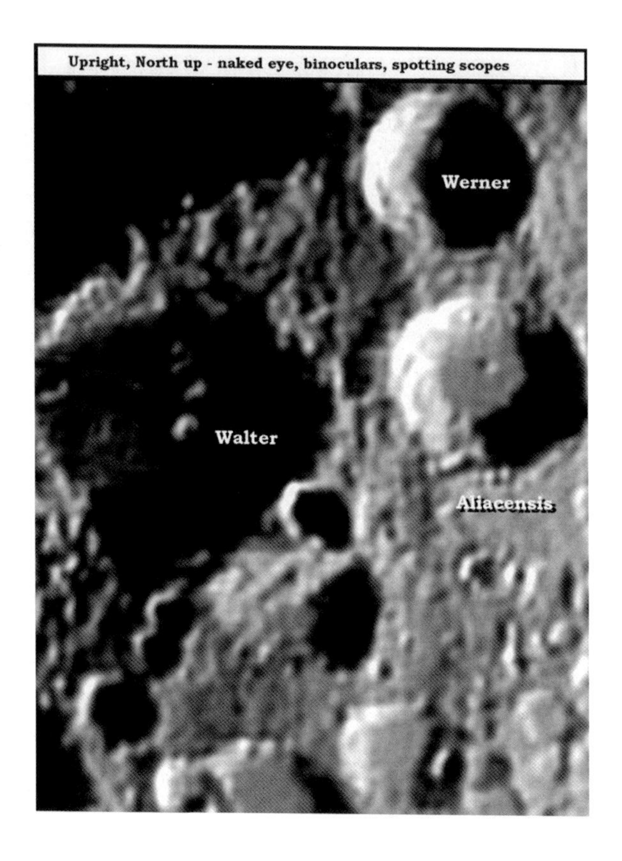
Upright, North up - naked eye, binoculars, spotting scopes
Werner
Walter
Aliacensis

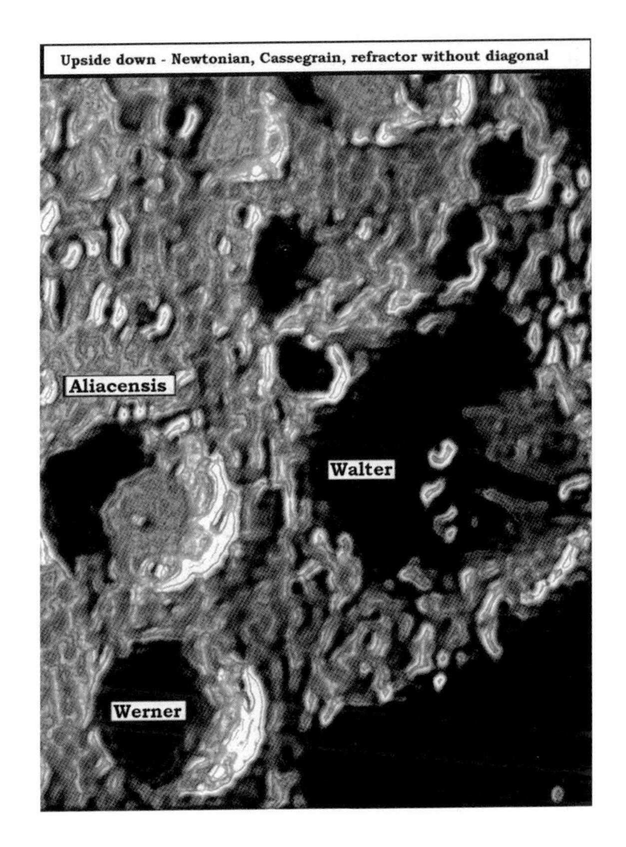
Upside down - Newtonian, Cassegrain, refractor without diagonal
Aliacensis
Walter
Werner

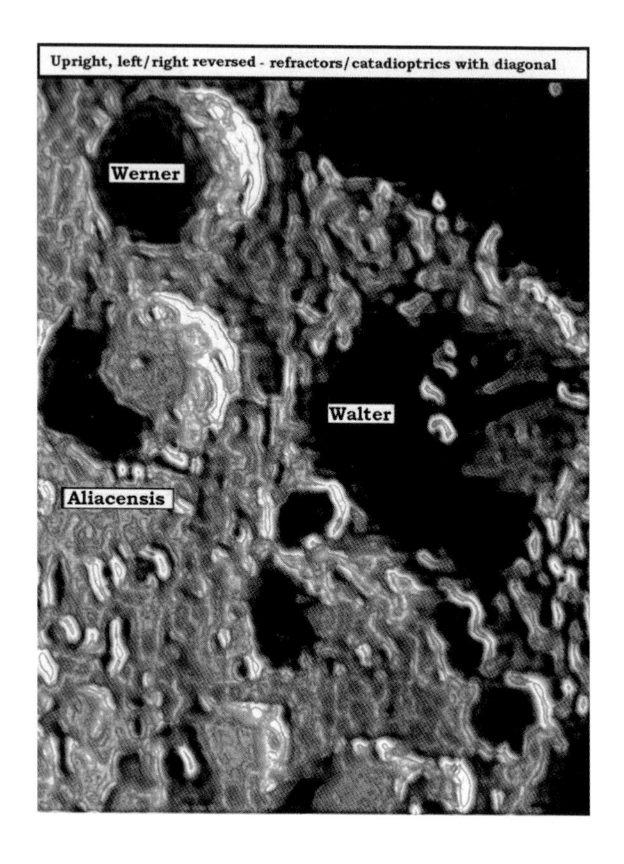
Upright, left/right reversed - refractors/catadioptrics with diagonal
Werner
Walter
Aliacensis

## CHART 32

# MORETUS

Moretus (70 miles) is a great crater that is not too far from the lunar South Pole. It is one of the deepest craters and has terraced walls that rise at least 2½ miles above the floor. There is a great central mountain mass that cast a long shadow as the sun rises on it. My attempt to determine the height of this mountain results in a height of about 1½ miles. This is close to other published results. This makes it the highest central peak I have observed. At first quarter you can easily recognize this formation. Be sure to point it out to your guests and friends.

Short (30 miles) is just south of Moretus. It is very deep crater, maybe deeper than Moretus.

Curtius (58 miles) is northeast of Moretus. The floor is flat except in the northwest where it is rough and has hills and craters.

Cysatus (30 miles) is another crater that is more than 2 miles deep. It remains in deep shadow while most of the other craters in the immediate area have much smaller shadows

Gruemberger (57 miles) is a battered formation. It may be about 2 miles deep. The floor is rough and there is a deep 12 mile crater near the western rim. Moretus is a distinctive formation and once you locate it you will easily find its neighbors.

D. Spain, *The Six-Inch Lunar Atlas: A Pocket Field Guide*,
DOI 10.1007/978-0-387-87610-8_34, 

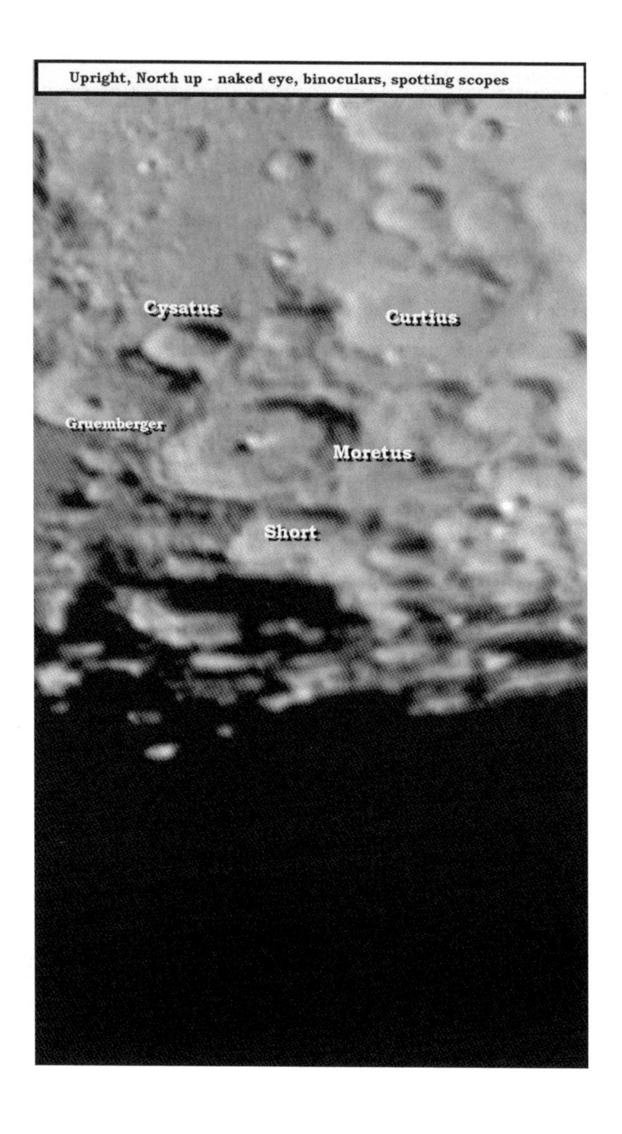
Upright, North up - naked eye, binoculars, spotting scopes
Cysatus
Curtius
Gruemberger
Moretus
Short

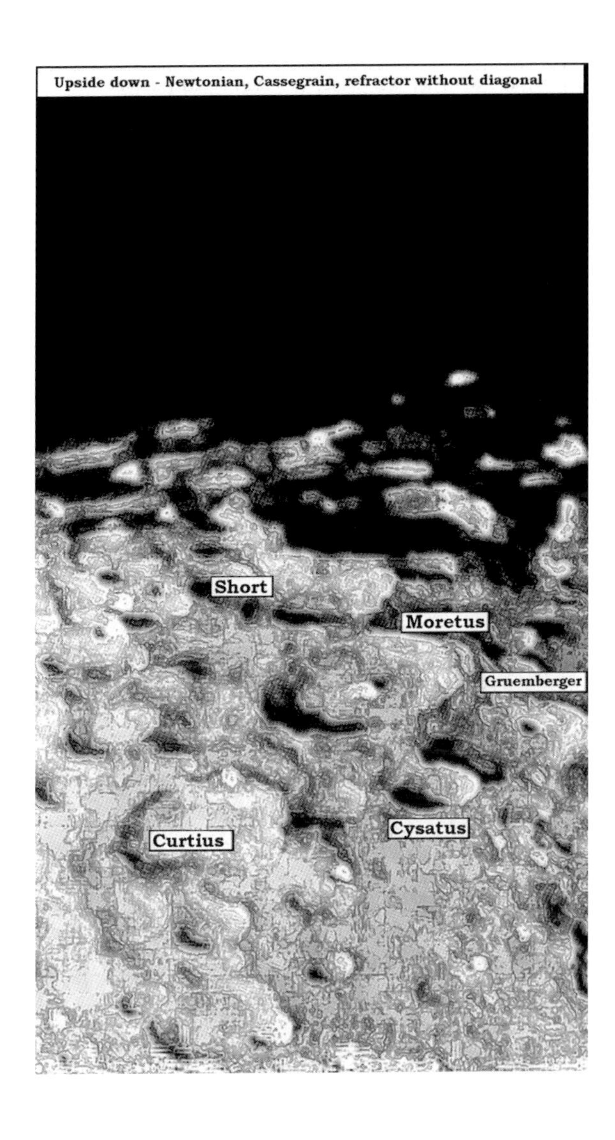
Upside down - Newtonian, Cassegrain, refractor without diagonal
Short
Moretus
Gruemberger
Curtius
Cysatus

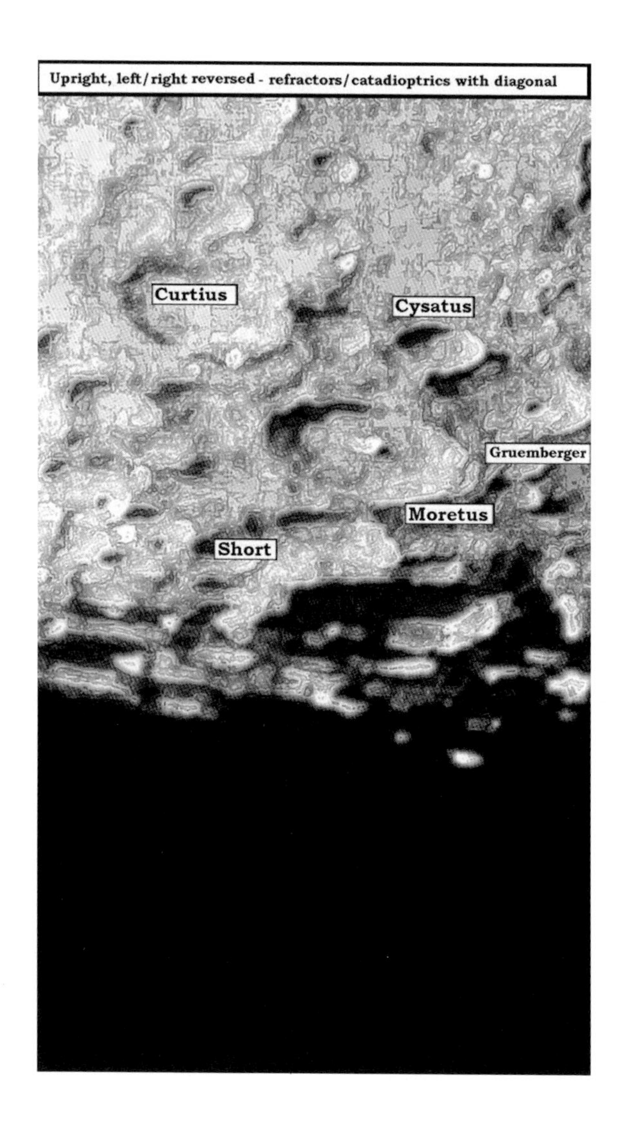
Upright, left/right reversed - refractors/catadioptrics with diagonal
Curtius
Cysatus
Gruemberger
Moretus
Short

CHART 33

# LALANDE

Lalande (15 miles) is a sharp rimmed crater that is about 1½ mile deep. There is a very small central hill. Notice the small block of hills, about 20 miles long just to the southwest. They are about ½ mile high. Also, notice the prominent isolated peak to the southwest of the hills. I would estimate it to be about ¾ mile high. The best time to observe this region is 1 day after first quarter.

Davy (20 miles) is a rough floored crater. Its walls are not very high, probably rising just over ½ mile above the floor. A pretty 9 mile wide crater that is about a mile deep is on the southeastern rim.

The highlands to the east are a region worth exploring under low illumination. In my 6 inch refractor I find valleys, craters, gouges and hills.

D. Spain, *The Six-Inch Lunar Atlas: A Pocket Field Guide*,
DOI 10.1007/978-0-387-87610-8_35, 

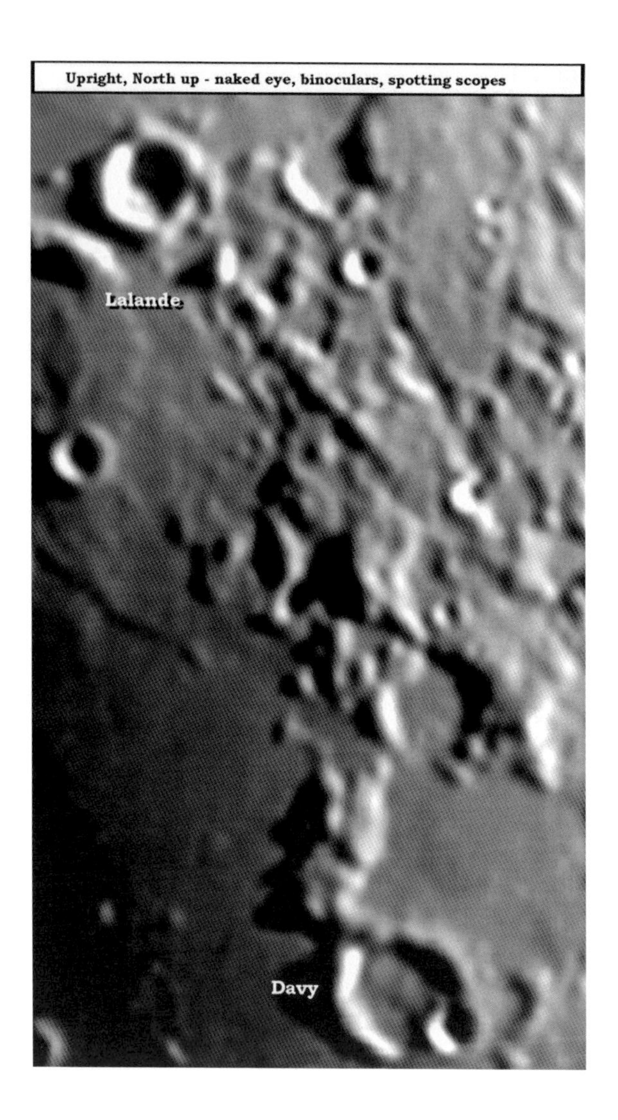
Upright, North up - naked eye, binoculars, spotting scopes
Lalande
Davy

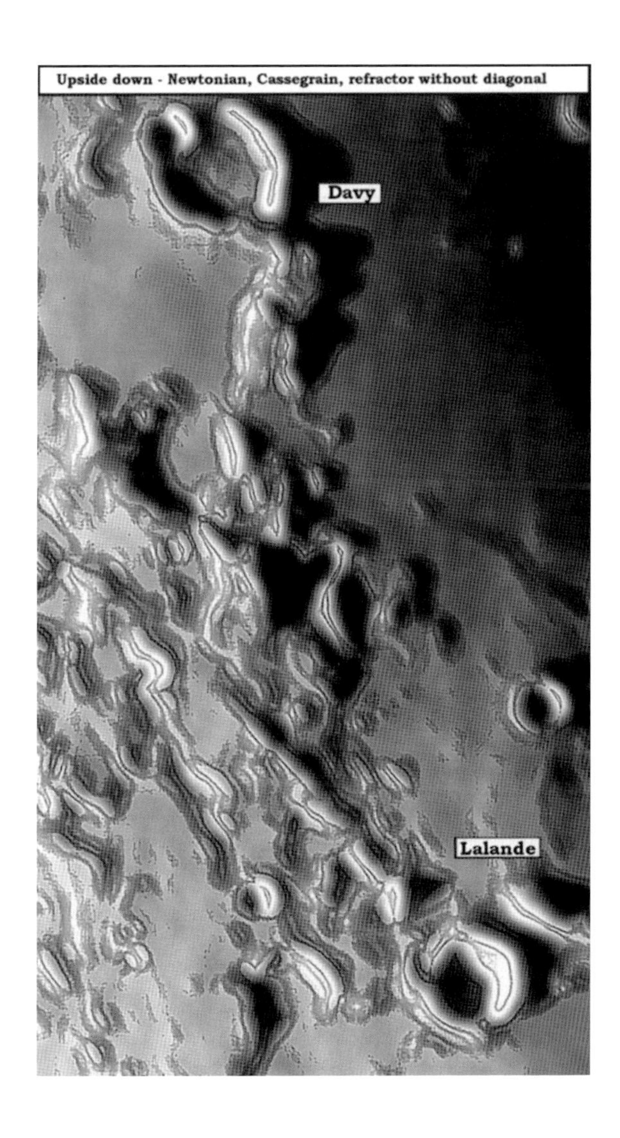
Upside down - Newtonian, Cassegrain, refractor without diagonal
Davy
Lalande

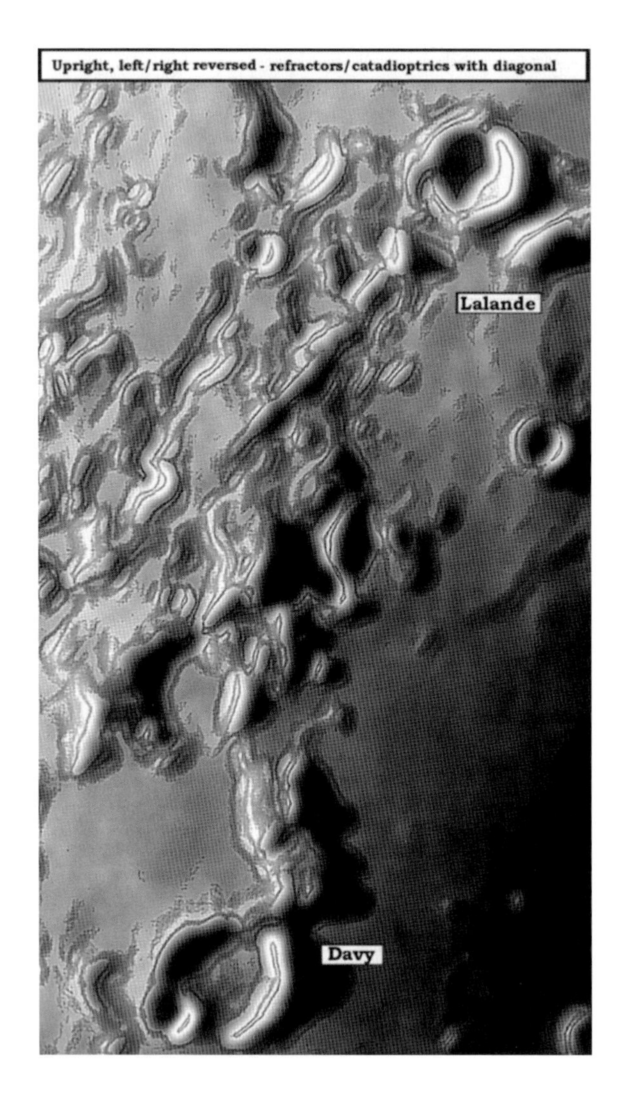
Upright, left/right reversed - refractors/catadioptrics with diagonal
Lalande
Davy

CHART 34

# RUPES RECTA

Rupes Recta is the most famous fault on the lunar surface. Known commonly as the Straight Wall, it is where the surface has cracked and slipped below the eastern surface. It is about 70 miles in length and ends near a 3 mile crater at its northern terminus. At its southern end there is a grouping of hills. It doesn't take a lot of imagination to visualize these hills as the handle and hilt of a sword and the Straight Wall as the blade of the sword. The best time to see this fascinating fault is 1 day after first quarter. It is then easily visible in my 2.4 inch refractor. It looks like the fault is a sheer drop, but in reality it slopes downward. Published estimates give a range from 40° to 7°. I am inclined to believe that it slopes no more than a gentle 10°. My personal estimate of the height is 800–1,200 feet. This formation is a must show at any public observation and to a beginning lunar observer.

Thebit (35 miles) is the large crater east of the Straight Wall. It is a deep crater with terraced walls that are almost 2 miles high. There is a very prominent 12 mile crater on its western rim. That crater in turn has a 7 mile crater on its western rim. Thebit's floor is rough with hills and crests.

Birt (10 miles) is a sharp rimmed crater to the west of the Straight Wall. It is about 2 miles deep and has a prominent little 4 mile crater on its eastern rim. There is a rill that starts just to the west of Birt. It makes a gentle curve to the north for about 30 miles. I have on rare occasion glimpsed it in my 4 inch refractor under very good conditions. It is an easy target find with the 6 inch.

This entire region will lure you back time and time again. I can guarantee that it will become one of your top 10 favorite areas on the lunar surface.

D. Spain, *The Six-Inch Lunar Atlas: A Pocket Field Guide*,
DOI 10.1007/978-0-387-87610-8_36, 

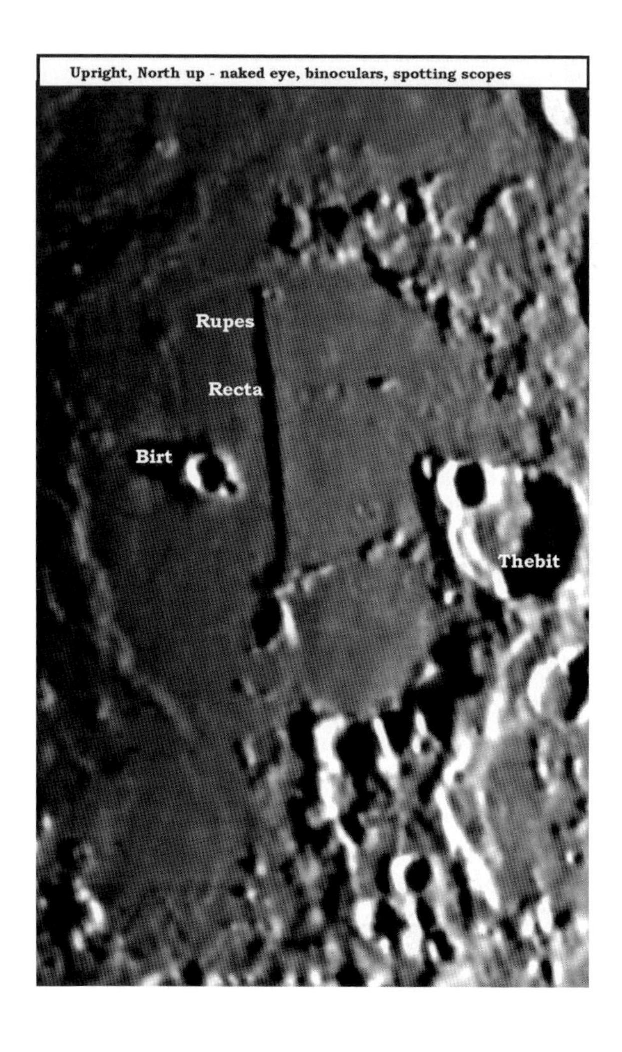
Upright, North up - naked eye, binoculars, spotting scopes
Rupes
Recta
Birt
Thebit

Upside down - Newtonian, Cassegrain, refractor without diagonal
Thebit
Birt
Rupes
Recta

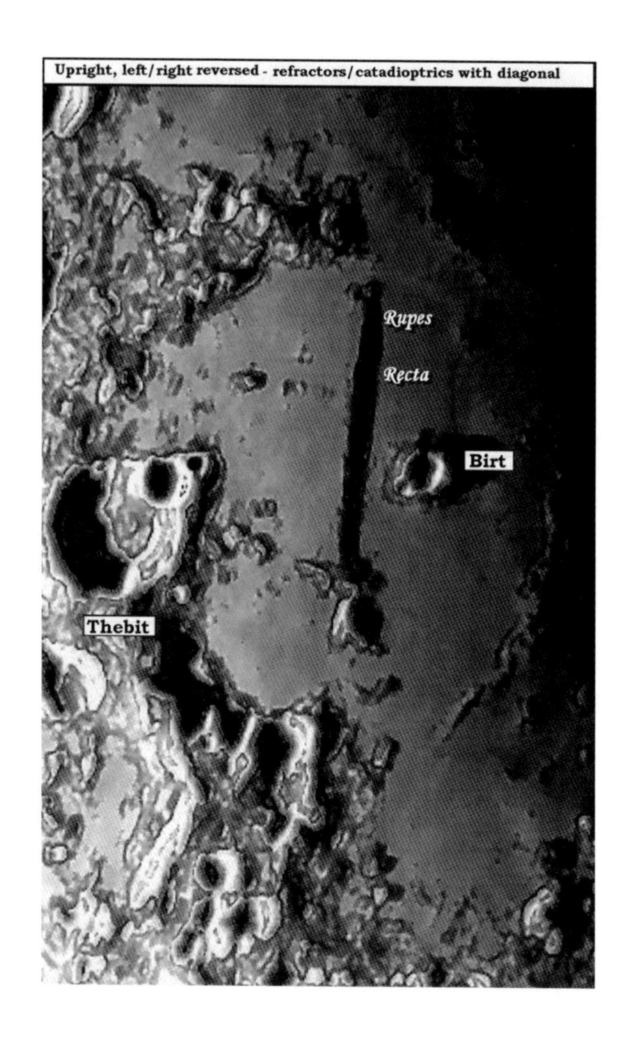
Upright, left/right reversed - refractors/catadioptrics with diagonal
Rupes
Recta
Birt
Thebit

## CHART 35

# DESLANDRES

Deslandres (142 miles) is a great ruined formation. To my eye it is in the shape of a great square. The walls are greatly eroded by later impacts and its great flat floor is the home of craters from 38 miles across to just craterlets and pits. There is a nice chain of craters on the northeast floor. The sun rises on Deslandres at first quarter and under its grazing rays there is great detail to see here, even in small scopes.

Lexell (38 miles) is partially ruined and part of its northern wall is open to the great plain of Deslandres. Its floor is rough and there is a nice central peak. Look for it on the southeastern floor of Deslandres.

Ball (25 miles) is a well formed crater with terraced walls and over 1½ miles deep. There is a massive central mountain over ½ mile high.

Hell (20 miles) is on Deslandres' western floor. It is just over 1 mile deep and has a large central mountain that is not quite ½ mile high.

Take the time to thoroughly explore this giant crater and its children. You will see something new each time, from a crater you missed before to a new hill popping out of the shadows.

D. Spain, *The Six-Inch Lunar Atlas: A Pocket Field Guide*,
DOI 10.1007/978-0-387-87610-8_37, 

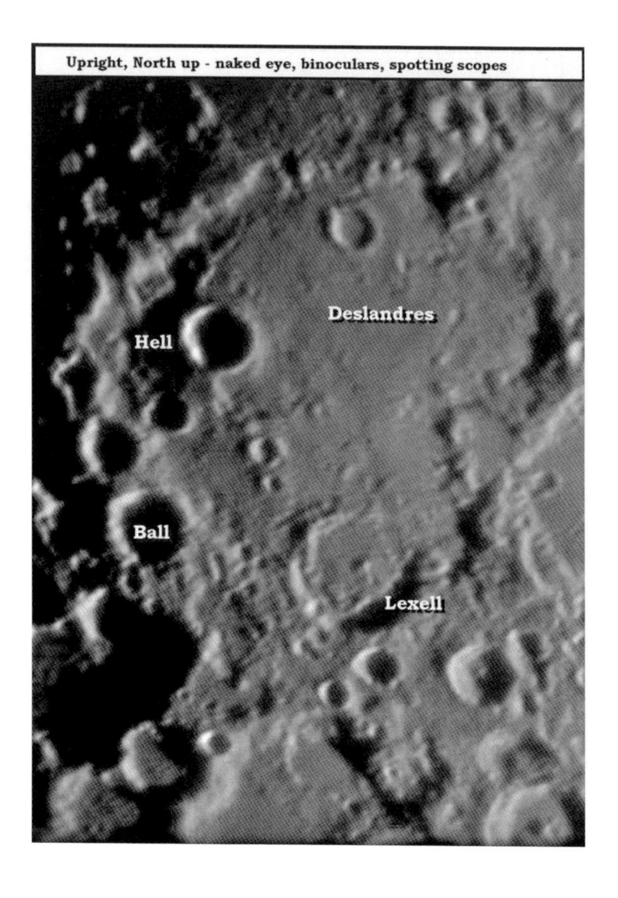
Upright, North up - naked eye, binoculars, spotting scopes
Deslandres
Hell
Ball
Lexell

Upside down - Newtonian, Cassegrain, refractor without diagonal
Lexell
Ball
Deslandres
Hell

Upright, left/right reversed - refractors/catadioptrics with diagonal
Deslandres
Hell
Ball
Lexell

## CHART 36

# TYCHO

Tycho (52 miles) is one of the youngest large craters on the Moon. Its walls are heavily terraced and rise upwards of 2¾ miles above the floor. There is a large central mountain mass that reaches up to about 1 mile from its base. Around the floor there are small hills scattered here and there. While Tycho is located in an area of heavy cratering it is so prominent that it is easy to spot 1 day after first quarter. This beautiful crater is another must show formation at any public observation.

Street (35 miles) is a fairly shallow crater that is less than 1 mile deep. There is some terracing of its walls that surround a flat floor. I see little detail on the floor, even in my 6 inch refractor.

Pictet (38 miles) has walls that are about 1½ miles high. The flat floor does reveal some low hills.

D. Spain, *The Six-Inch Lunar Atlas: A Pocket Field Guide*,
DOI 10.1007/978-0-387-87610-8_38, 

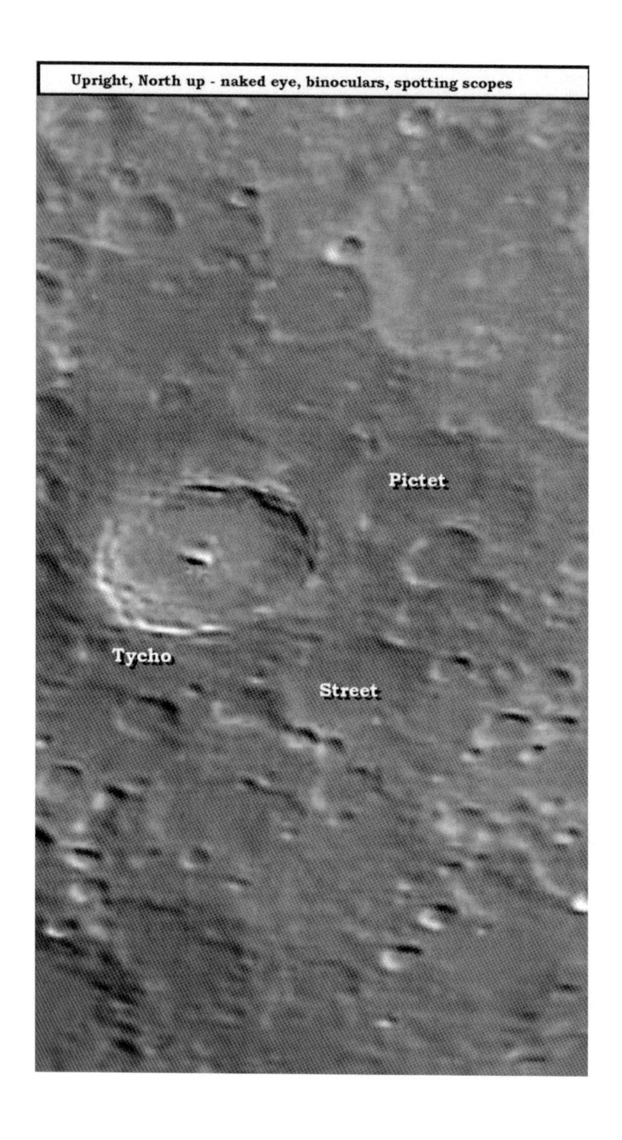
Upright, North up - naked eye, binoculars, spotting scopes
Pictet
Tycho
Street

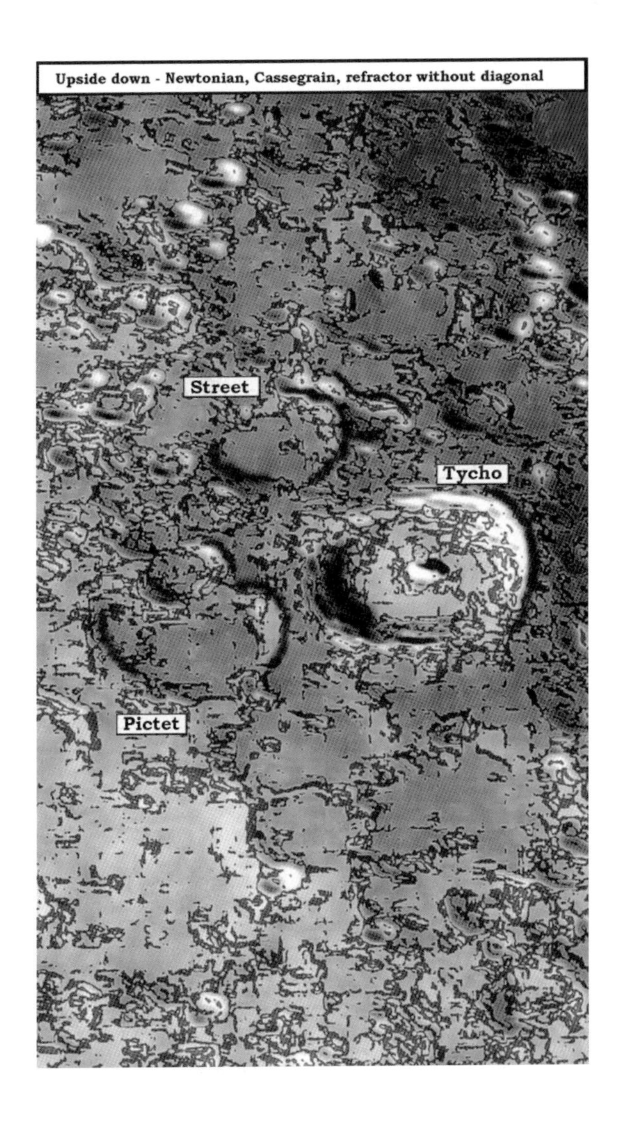
Upside down - Newtonian, Cassegrain, refractor without diagonal
Street
Tycho
Pictet

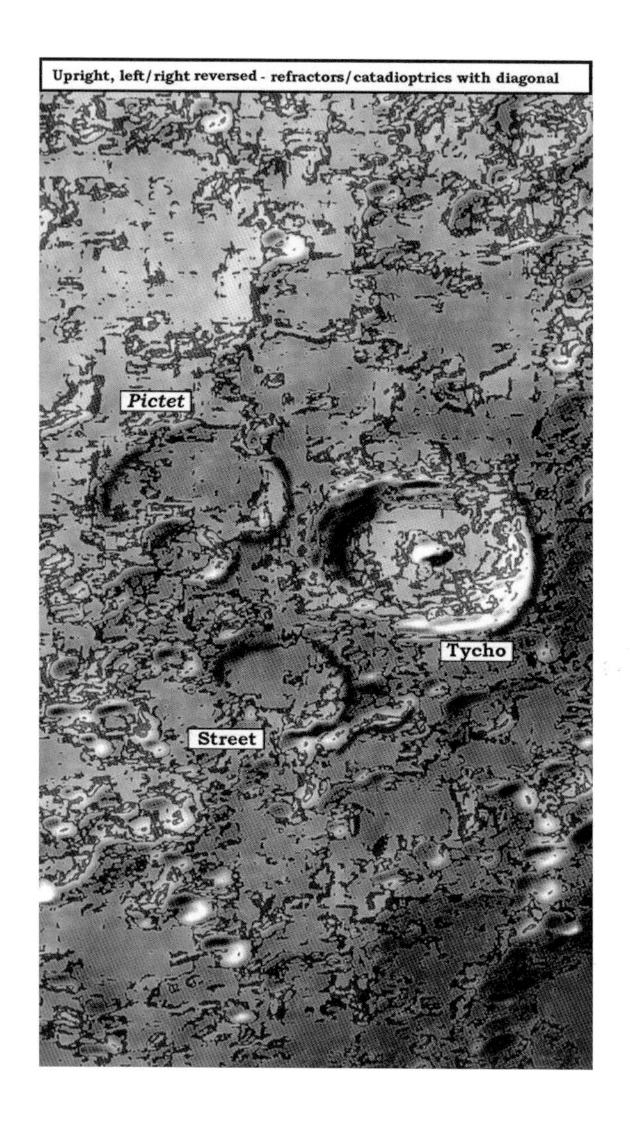
Upright, left/right reversed - refractors/catadioptrics with diagonal
Pictet
Tycho
Street

## CHART 37

# CLAVIUS

Clavius (136 miles) is a very impressive lunar giant. Because of its huge size and great blocky walls it is impossible to miss 1 day after first quarter. The walls must be at least 2 miles about the vast flat floor. In addition to the two large craters, Porter on the northeastern rim and Rutherfurd on the southeast, the floor is strewn with dozens of small craters. There is a series of good size craters that run through the center of the floor in an east-west direction. Five of this line of craters are easily visible in my 2.4 inch refractor. There is a grouping of small hills near the center of these craters that are easy to see in the 4 inch refractor. Look for other lines and crater chains on the floor. This is a visually beautiful formation that you must show off to anyone who visits your eyepiece.

Rutherfurd (32 miles) is about 1½ mile deep and abuts the southeastern wall of Clavius. There is a prominent off center peak rising from the rough and hilly floor. An impressive crater in its own right, it is overpowered by Clavius.

Porter (32 miles) is somewhat similar to Rutherfurd is size and depth. It also has a prominent central peak.

Maginus (100 miles) is Clavius' smaller brother to its northeast. It is worth the effort to leave Clavius and explore. The walls are not as high and they are heavily cratered, especially on the north and west rim. There is a small central hill.

D. Spain, *The Six-Inch Lunar Atlas: A Pocket Field Guide*,
DOI 10.1007/978-0-387-87610-8_39, 

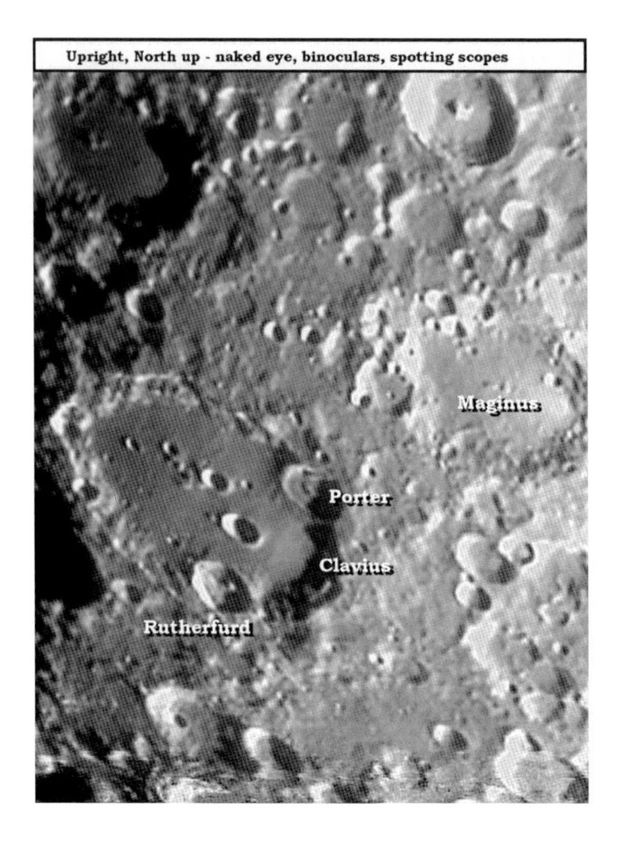
Upright, North up - naked eye, binoculars, spotting scopes
Maginus
Porter
Clavius
Rutherfurd

Upside down - Newtonian, Cassegrain, refractor without diagonal
Rutherfurd
Clavius
Porter
Maginus

Upright, left/right reversed - refractors/catadioptrics with diagonal
Maginus
Porter
Clavius
Rutherfurd

## CHART 38

# FRA MAURO

Fra Mauro (58 miles) is a low walled formation. Look for it 1 day after first quarter. A low sun angle is needed to view the rim surrounding its large flat floor. A large section of the eastern rim is entirely missing. There are some shallow rills that I can make out in my 6 inch refractor at lunar sunrise in this area. It is conjoined with Parry and Bonpland on its southern rim. When viewed through a telescope that inverts the image this trio reminds me of Mickey Mouse.

Bonpland (36 miles) is another ruined formation. The walls are high only where its rim interacts with Fra Mauro and Parry on its north and eastern rim. Its southern rim is marked by an outline of small hills. The flat floor does contain shallow rills and many small craterlets that will require at least 6 inches of aperture to spot.

Parry (29 miles) is the best formed of the trio with a complete, but low rim. Like its partners, it has a couple of shallow rills on its flat floor.

Tolansky (8 miles) is a small, but prominent crater, just below the above trio. It is about ½ mile deep with a flat floor.

Guericke (35 miles) is yet another low wall ruined crater in this region. The rim at its highest is probably less ½ mile above the floor. It has a nice 4 mile crater on its southwestern floor.

This area is very interesting under grazing light. If you have at least 4 inches of aperture you will be well rewarded by studying this region.

D. Spain, *The Six-Inch Lunar Atlas: A Pocket Field Guide*,
DOI 10.1007/978-0-387-87610-8_40, 

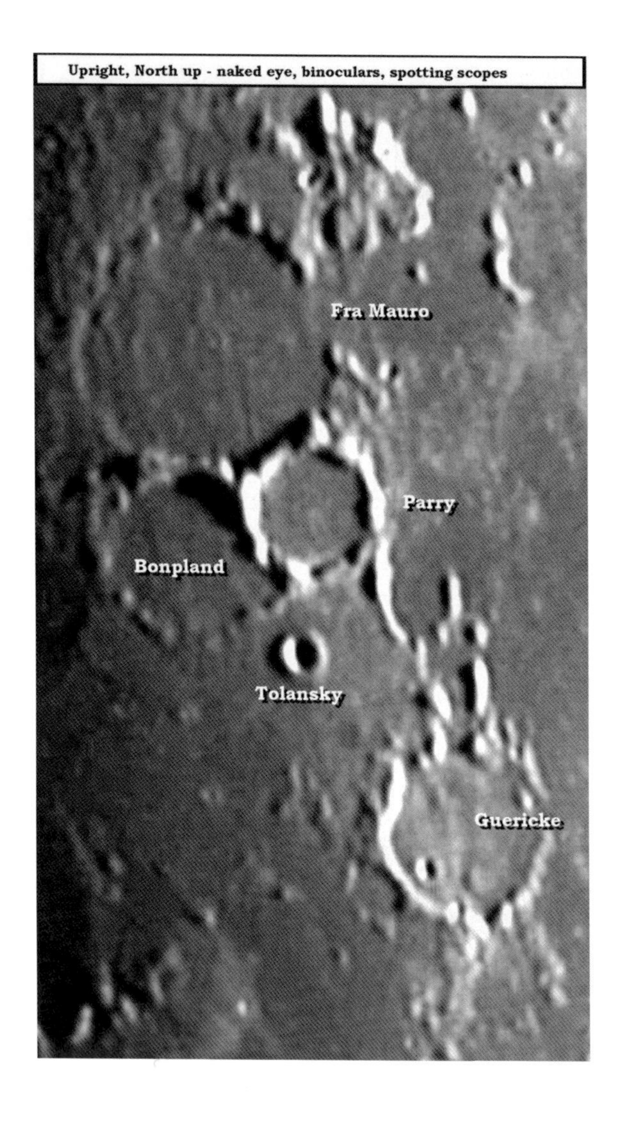
Upright, North up - naked eye, binoculars, spotting scopes
Fra Mauro
Parry
Bonpland
Tolansky
Guericke

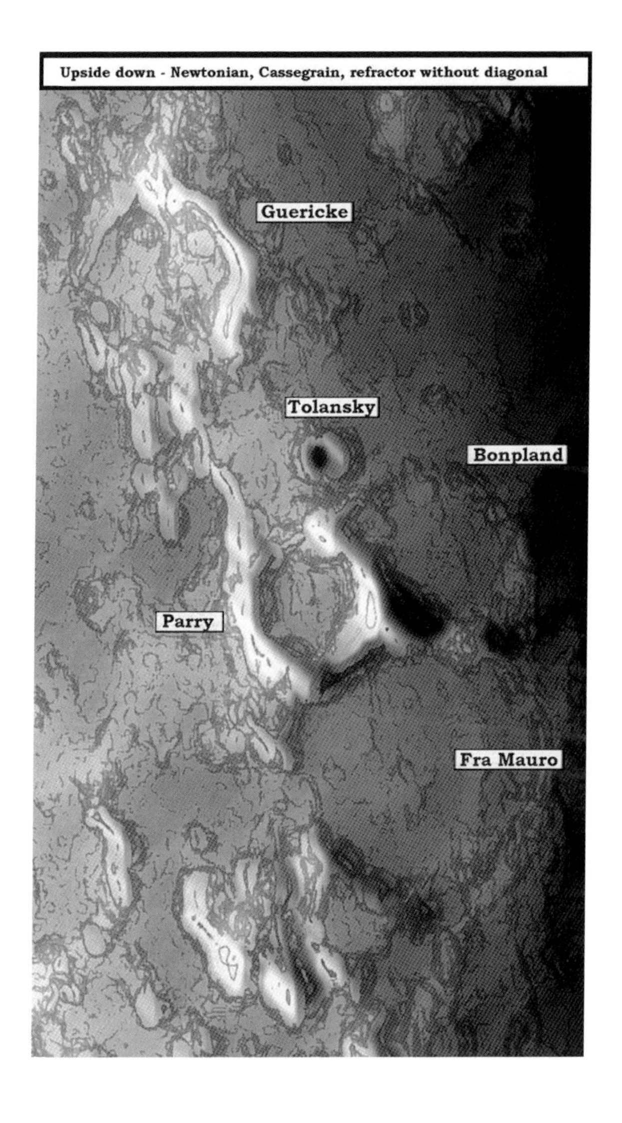
Upside down - Newtonian, Cassegrain, refractor without diagonal
Guericke
Tolansky
Bonpland
Parry
Fra Mauro

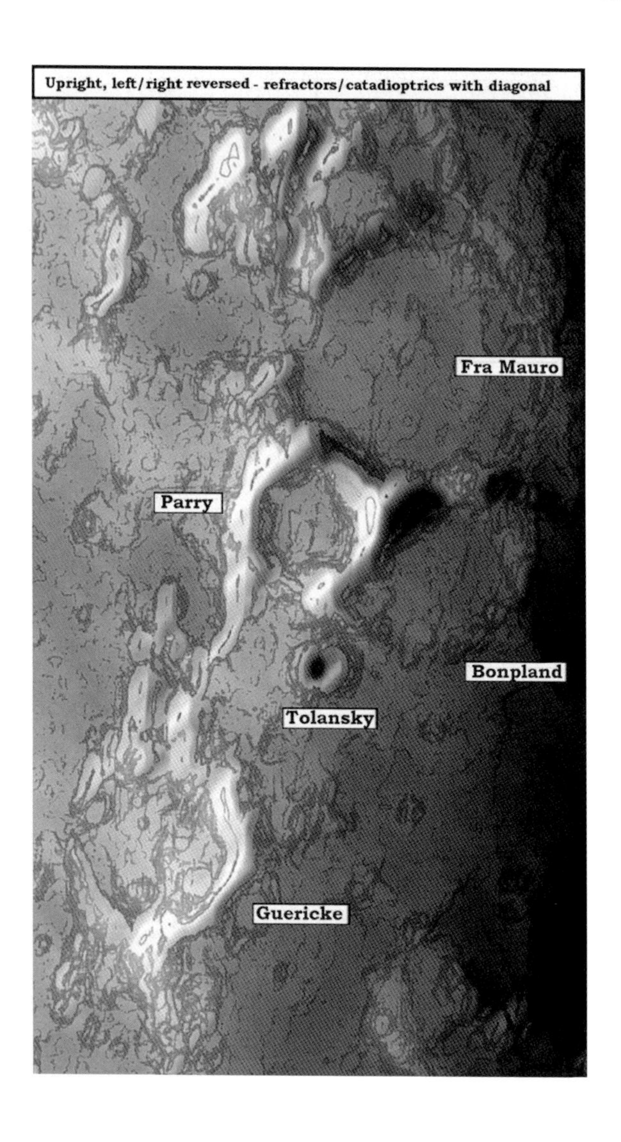
Upright, left/right reversed - refractors/catadioptrics with diagonal
Fra Mauro
Parry
Bonpland
Tolansky
Guericke

## CHART 39

# PITATUS

Pitatus (62 miles) is a low rim crater with a dark floor. There is an open pass to Hesiodus on its western rim. When the sun rises on Pitatus 1 day after first quarter a small central peak, about ½ mile high, becomes visible. My 6 inch refractor reveals a series of rills that parallel the inside walls.

Hesiodus (26 miles) is a battered old crater with low walls. There is a nice little 3 mile crater near the center of the floor.

Wurzelbauer (53 miles) is a battered old crater with worn down walls. The floor is rough and is higher in the west.

Heinsius (39 miles) is a crater that is lucky to have survived the impacts that created three craters at least 12 miles across on its southern floor and rim. Its surviving walls are 1½ miles about the floor.

Cichus (25 miles) is a sharp rimmed crater 1½ mile deep. It has a nice 7 mile crater on its southwestern rim.

Weiss (40 miles) is missing its northern wall and is now really a bay. There is a 10 mile crater on its northern floor that reminds me of a fort guarding the entrance to the bay.

D. Spain, *The Six-Inch Lunar Atlas: A Pocket Field Guide*,
DOI 10.1007/978-0-387-87610-8_41, © Springer Science+Business Media, LLC 2009

Upright, North up - naked eye, binoculars, spotting scopes
Hesiodus
Pitatus
Weiss
Cichus
Wurzelbauer
Heinsius

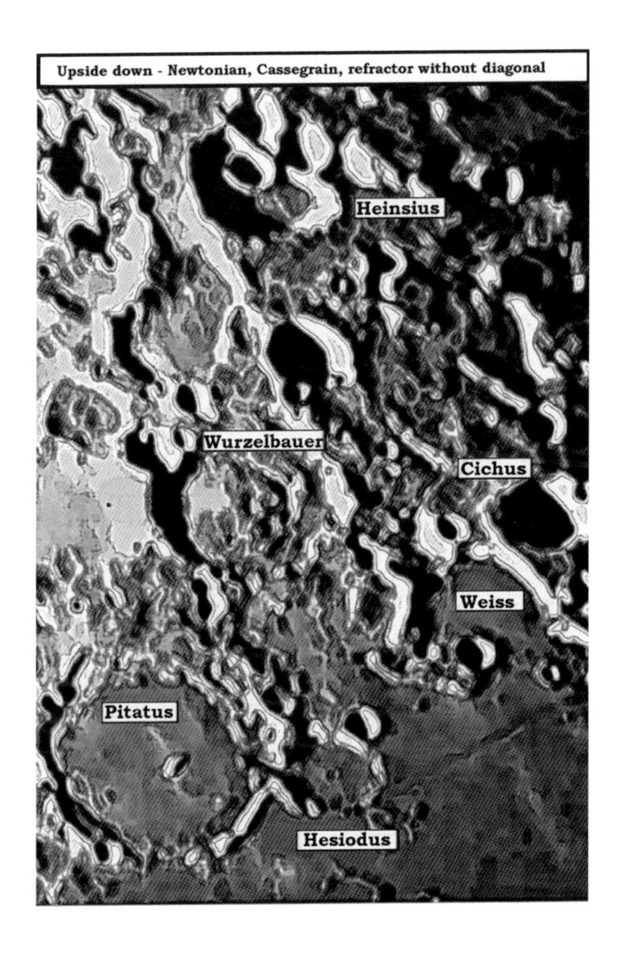
Upside down - Newtonian, Cassegrain, refractor without diagonal
Heinsius
Wurzelbauer
Cichus
Weiss
Pitatus
Hesiodus

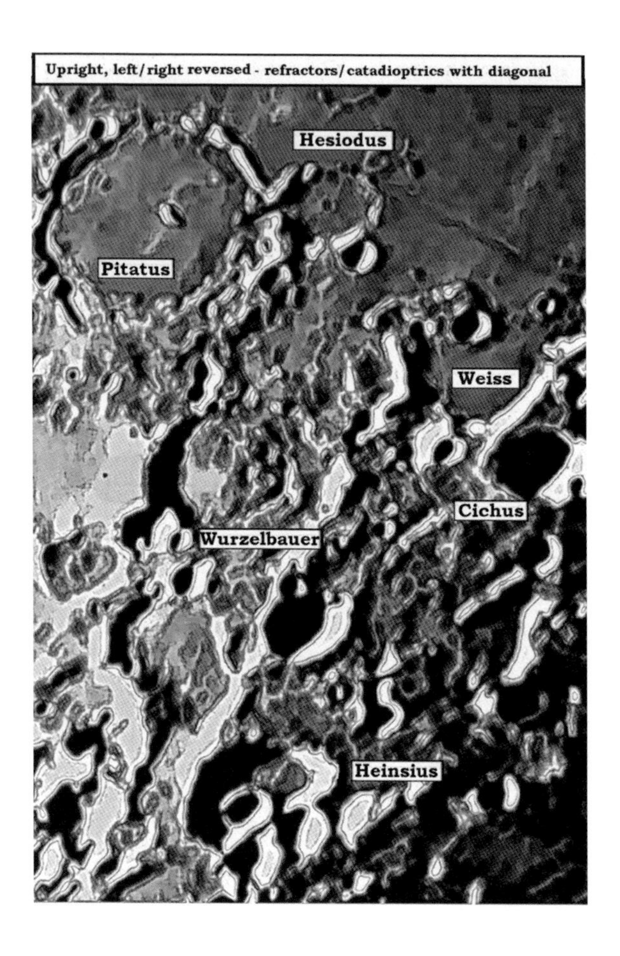
Upright, left/right reversed - refractors/catadioptrics with diagonal
Hesiodus
Pitatus
Weiss
Cichus
Wurzelbauer
Heinsius

## CHART 40

# LONGOMONTANUS

Longomontanus (88 miles) has terraced walls that are about 2 miles high. The photo, taken 2 days after first quarter, shows just the tops of the off center peaks. The great flat floor is dark and the northern rim is battered by several sizable craters.

Montanari (47 miles) has suffered from many impacts. The floor is flat to the north, but rough in the south.

Lagalla (52 miles) is all that remains of what must have an imposing crater. Its southern walls are over a mile high, but most of the rest of the formation is destroyed by the impact that formed Wilhelm.

Wilhelm (65 miles) is partially ruined by impacts, but its walls rise to about 2 miles in several places. Its floor is generally flat, but the eastern area is rough with craterlets and small hills.

While these formations are in the heavily cratered Southern Highlands, they are just west and south of Tycho, making them easy to locate. Also, Longomontanus is so large it is very easy to find. My 2.4 inch refractor shows a lot of detail here at 150×.

D. Spain, *The Six-Inch Lunar Atlas: A Pocket Field Guide*,
DOI 10.1007/978-0-387-87610-8_42, 

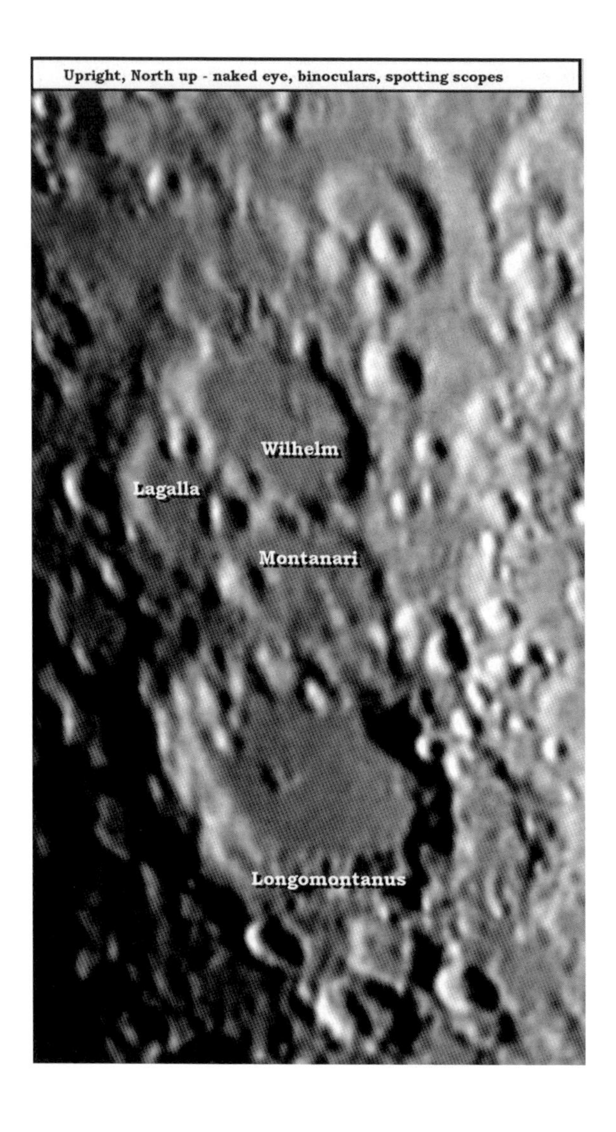
Upright, North up - naked eye, binoculars, spotting scopes
Wilhelm
Lagalla
Montanari
Longomontanus

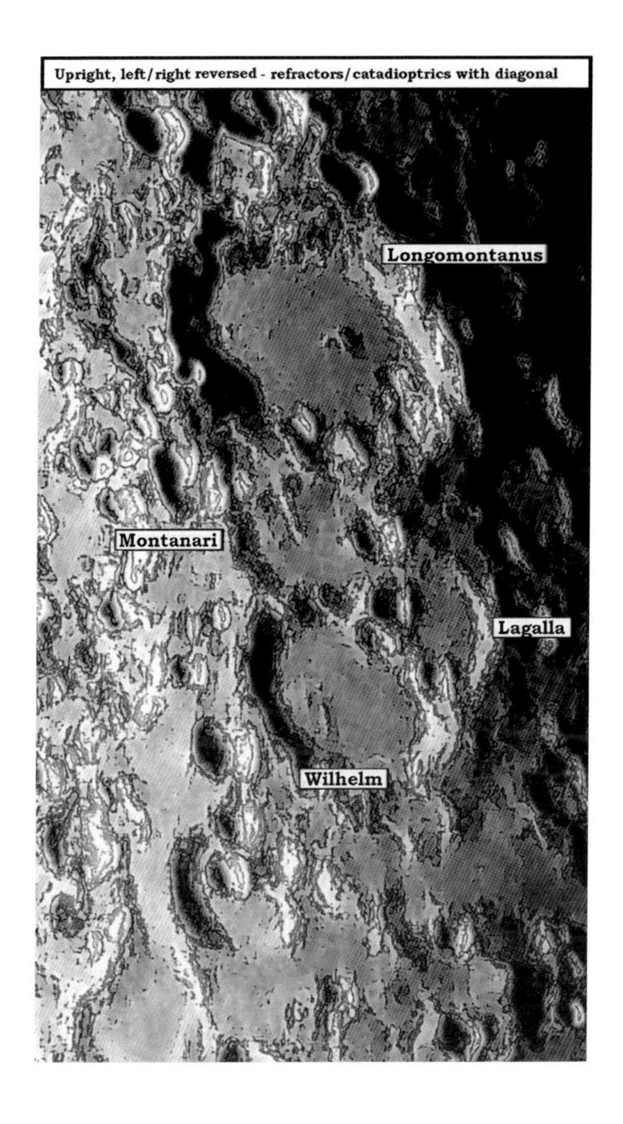
Upright, left/right reversed - refractors/catadioptrics with diagonal
Longomontanus
Montanari
Lagalla
Wilhelm

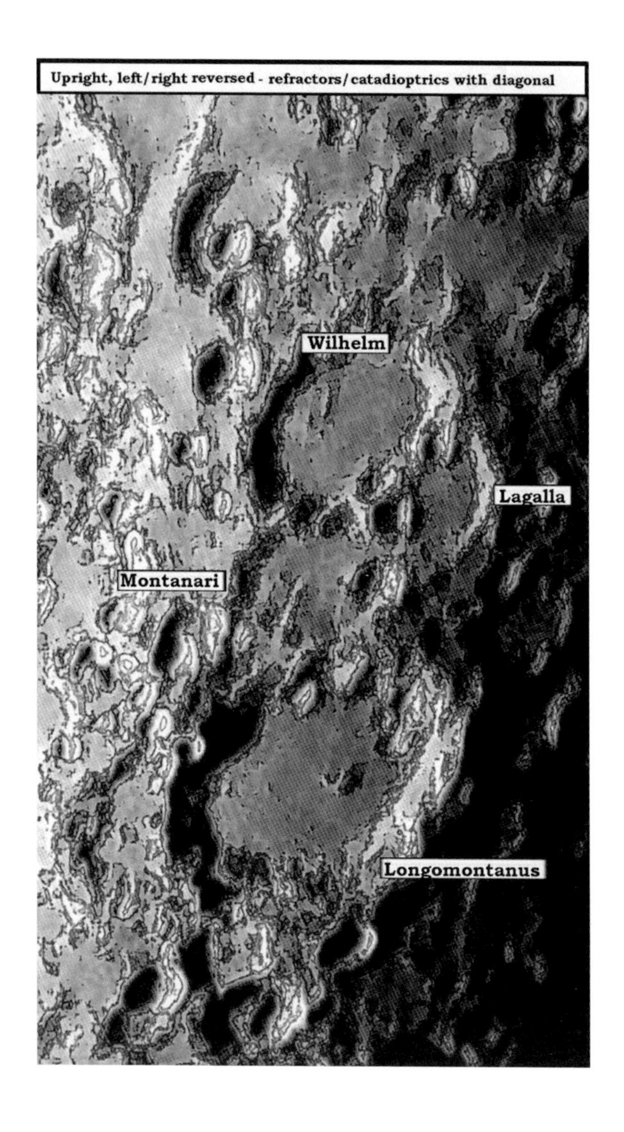
Upright, left/right reversed - refractors/catadioptrics with diagonal
Wilhelm
Lagalla
Montanari
Longomontanus

## CHART 41

# KIES

Kies (27 miles) is a mostly submerged crater with only a little of its rim still visible. The highest wall is only about ¼ mile high and a small section is completely missing from the western rim. The floor is flooded and the same dark color as the Mare Nubium, the Sea of Clouds that surround Kies. Under morning illumination 2 days after first quarter, this little formation and the area around it show great detail. Wrinkle ridges, small craters and rills are easily spotted.

Pi Kies (6 miles) is a typical large lunar dome. While I have seen it under very grazing sunlight with my 2.4 inch refractor, the 4 inch reveals it pretty easily. The 6 inch refractor will show a tiny crater pit in the center of the dome.

Rima Hesiodus is a long shallow rill. It starts at the northwestern corner of Hesiodus and runs southwest direction for about 180 miles. About 70 miles are visible on the photo. It is easy in my 6 inch refractor under low light.

D. Spain, *The Six-Inch Lunar Atlas: A Pocket Field Guide*,
DOI 10.1007/978-0-387-87610-8_43, 

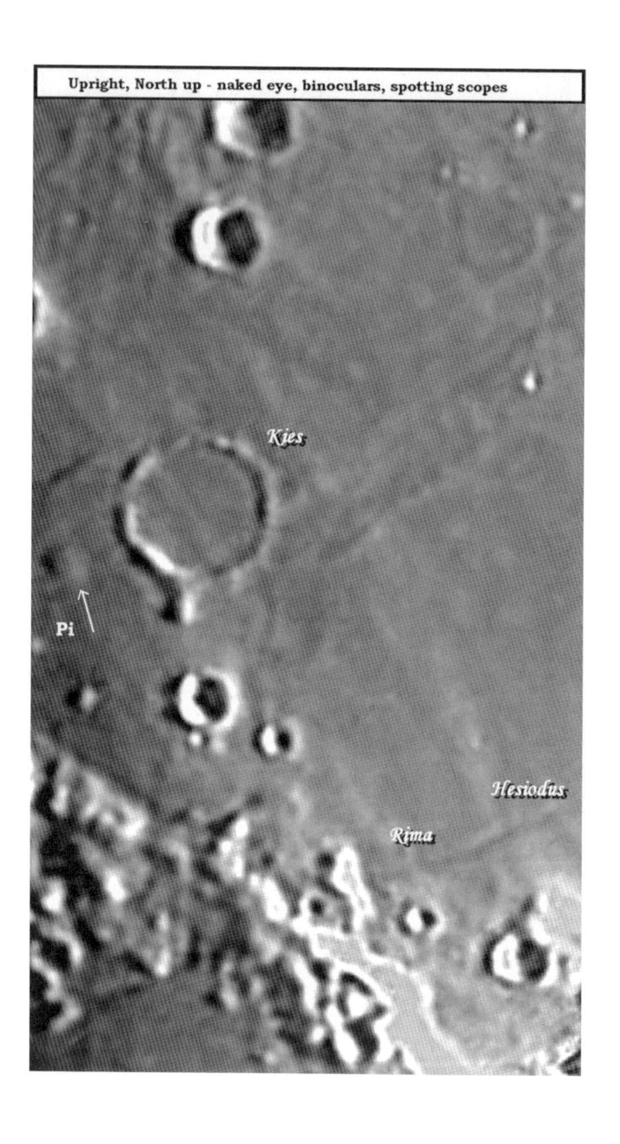
Upright, North up - naked eye, binoculars, spotting scopes
Kies
Pi
Hesiodus
Rima

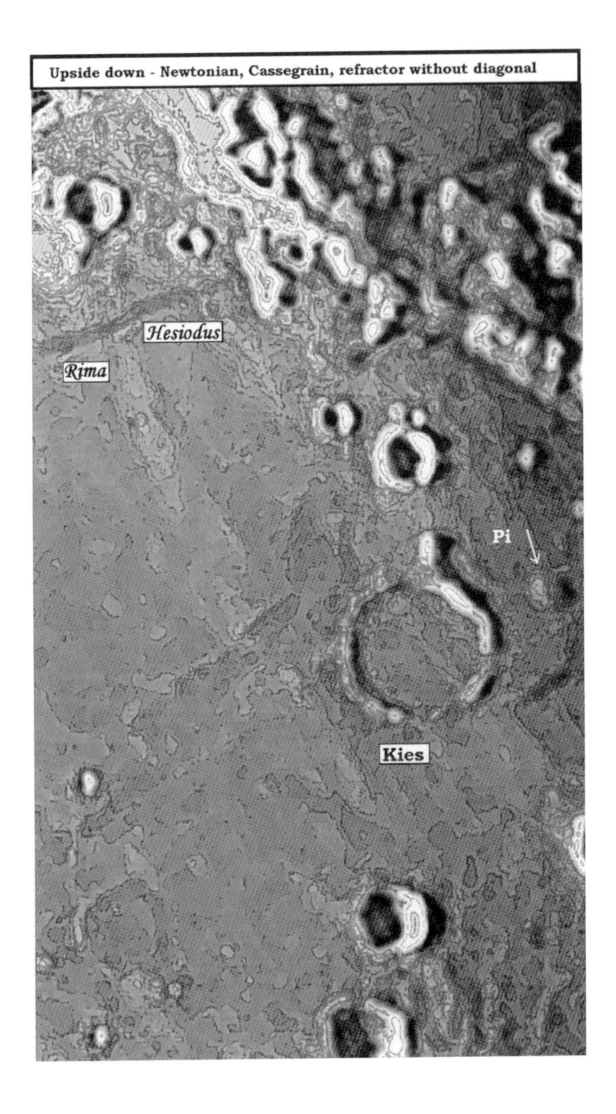
Upside down - Newtonian, Cassegrain, refractor without diagonal
Hesiodus
Rima
Pi
Kies

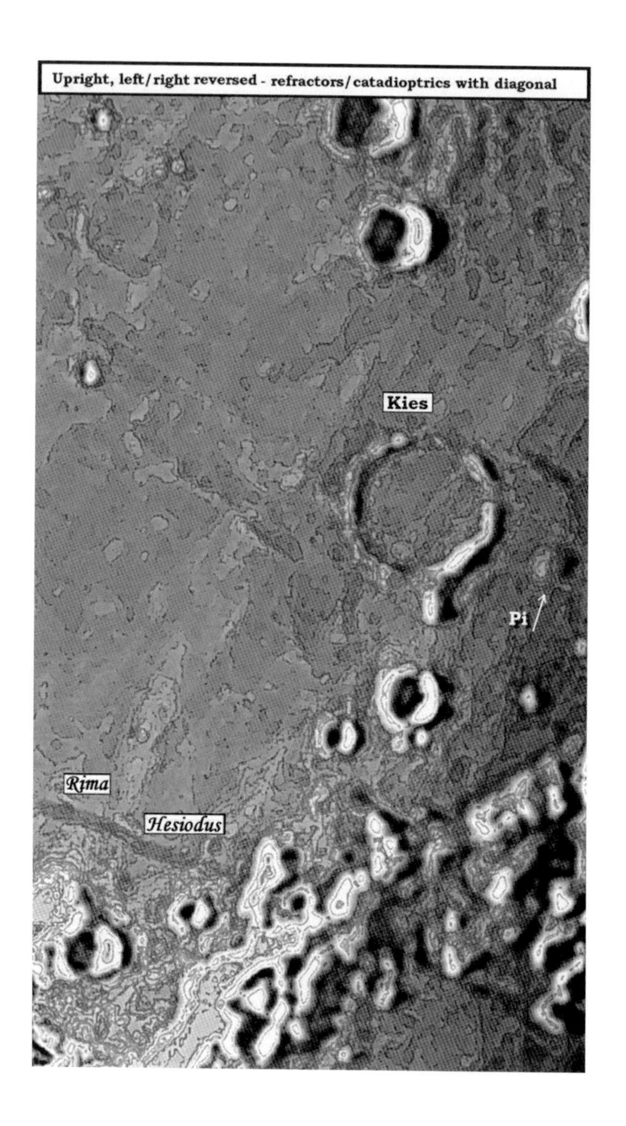
Upright, left/right reversed - refractors/catadioptrics with diagonal
Kies
Pi
Rima
Hesiodus

## CHART 42

# MONTES RECTI

Montes Recti is commonly known as the Straight Range. This little range of mountains is in the shape of a rectangle. They are about 55 miles long and 10 wide. The higher peaks reach up to 1 mile. Look for this peculiar stretch of mountains 2 days after first quarter. They are easily visible in small scopes.

Maupertuis (28 miles) is deformed, more square than circular and located in an upland area. It walls are probably about 1½ miles about the floor. The rough floor has ridges crossing it.

La Condamine (22 miles) is also in an upland region. Its walls reach probably to 1½ miles above the flat and rough floor.

This is an interesting area and it invites your inspection. Notice the wrinkle ridges out on the lava plain and the craters and hills in the upland area.

D. Spain, *The Six-Inch Lunar Atlas: A Pocket Field Guide*,
DOI 10.1007/978-0-387-87610-8_44, 

Upright, North up - naked eye, binoculars, spotting scopes
La Codamine
Montes Recti
Maupertuis

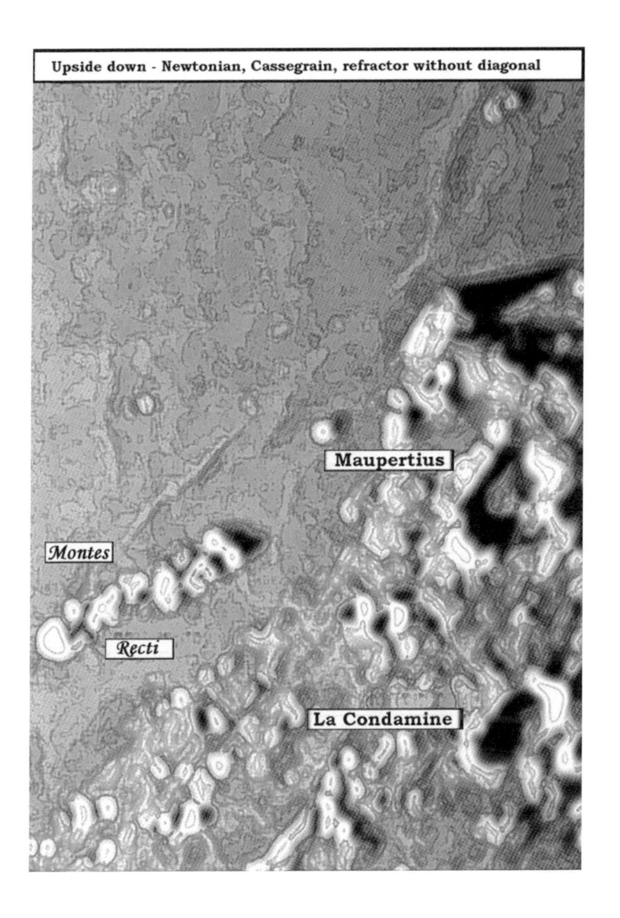
Upside down - Newtonian, Cassegrain, refractor without diagonal
Maupertius
Montes
Recti
La Condamine

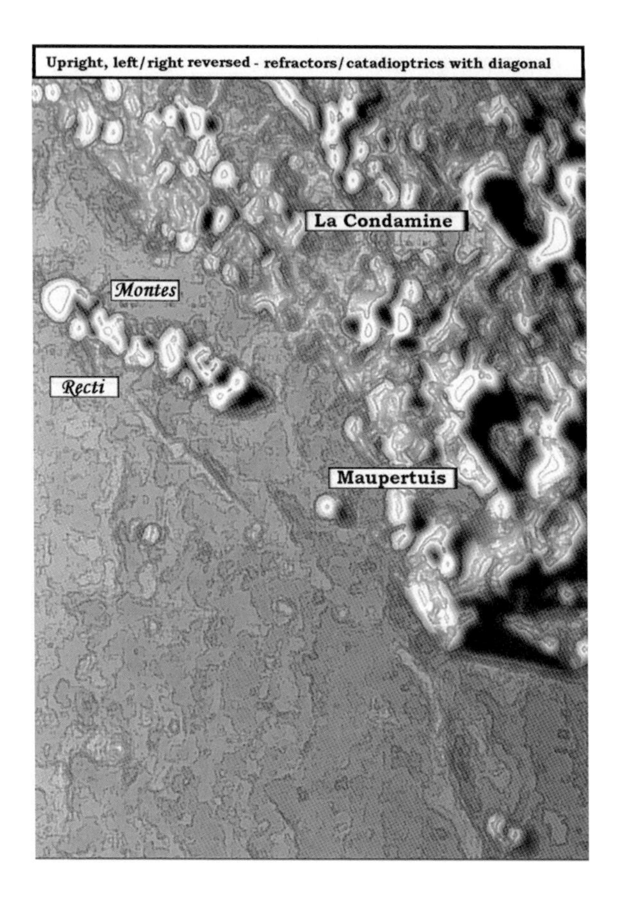
Upright, left/right reversed - refractors/catadioptrics with diagonal
La Condamine
Montes
Recti
Maupertuis

CHART 43

# HELICON

Helicon (15 miles) is out in the Mare Imbrium, the Sea of Rains. It has a sharp rim and is a little over 1 mile deep. It is well placed for observation 2 days after first quarter. The floor appears to be flat and featureless in my 6 inch refractor. It makes and attractive pair with Le Verrier, just 17 miles to the east.

Le Verrier (12 miles) is a little smaller than its western neighbor, but it is a little deeper. The flat floor reveals no detail in my 6 inch refractor.

Carlini (7 miles) is small, but very sharp rimmed crater. It is very deep for its size, with a depth of 1¼ mile.

McDonald (5 miles) is an even smaller, but still prominent crater out on the plain. It is somewhat over 1 mile deep.

Mons La Hire, La Hire's Mountain is an isolated massif. Easily visible in small scopes at lunar sunrise it rises about a mile above the plain. Notice the great wrinkle ridge to its northeast.

D. Spain, *The Six-Inch Lunar Atlas: A Pocket Field Guide*,
DOI 10.1007/978-0-387-87610-8_45, 

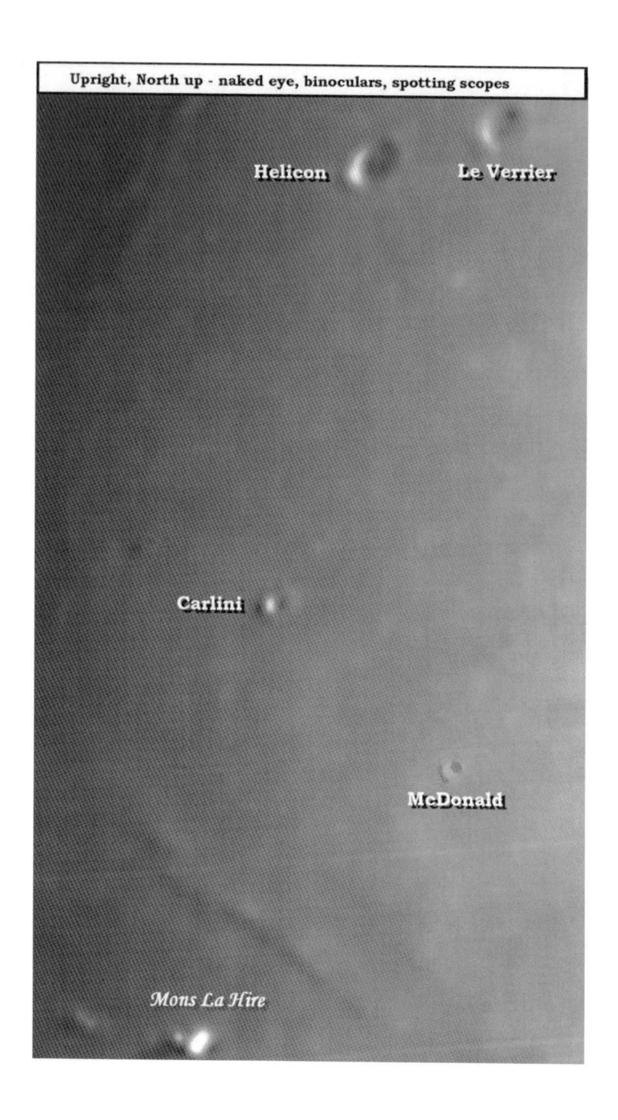
Upright, North up - naked eye, binoculars, spotting scopes
Helicon
Le Verrier
Carlini
McDonald
Mons La Hire

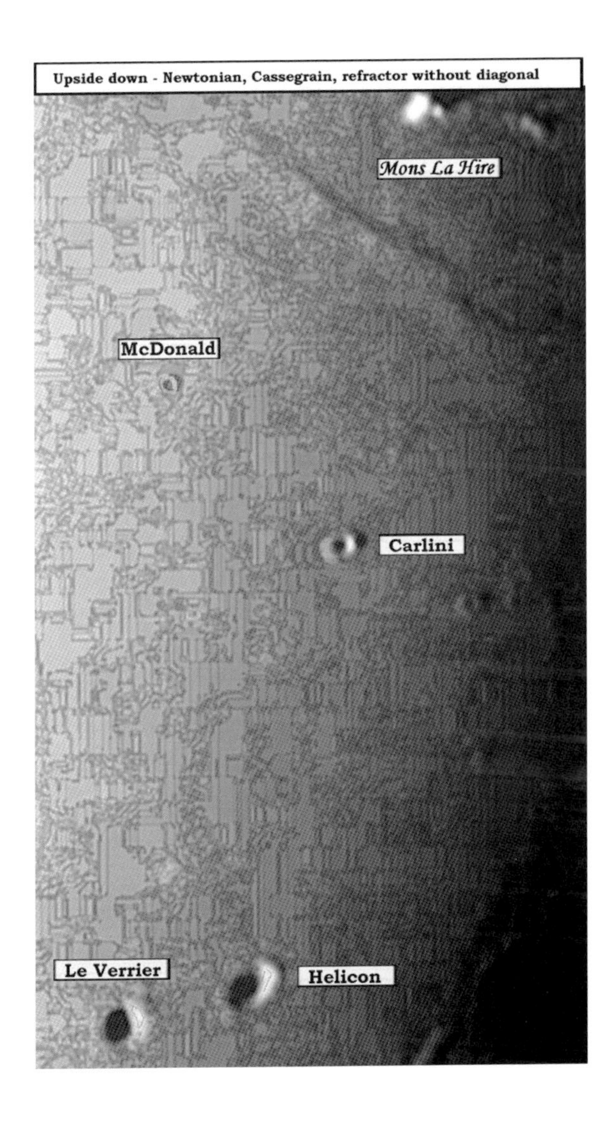
Upside down - Newtonian, Cassegrain, refractor without diagonal
Mons La Hire
McDonald
Carlini
Le Verrier
Helicon

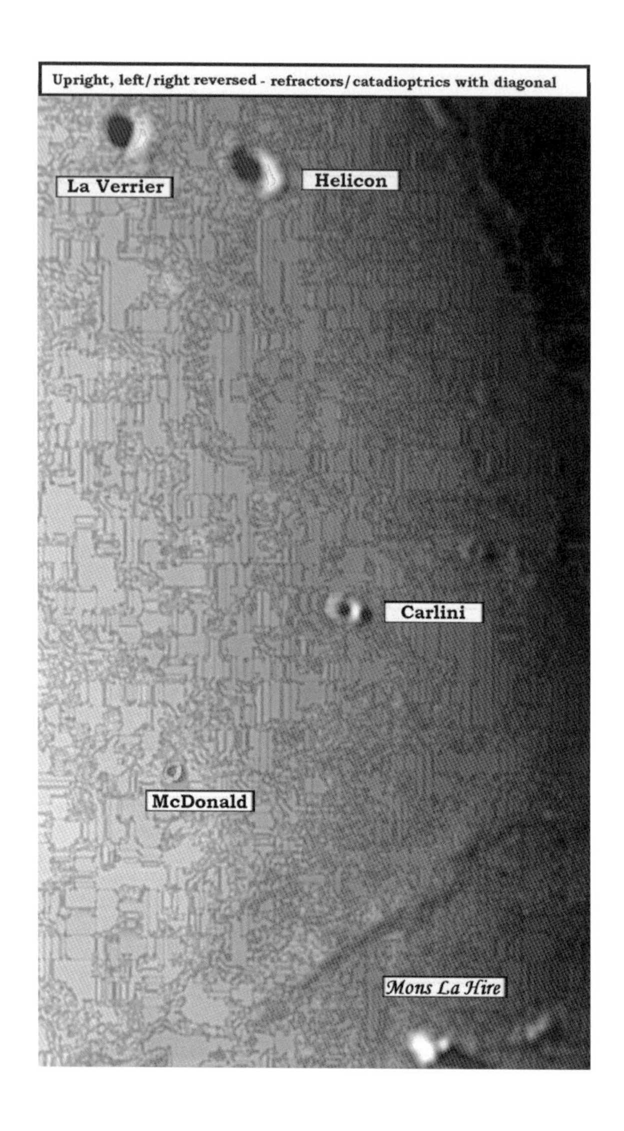
Upright, left/right reversed - refractors/catadioptrics with diagonal
La Verrier
Helicon
Carlini
McDonald
Mons La Hire

## CHART 44

# MONTES CARPATUS

Montes Carpatus or the Carpathian Mountains are a range that forms the southern shore of Mare Imbrium, the Sea of Rains. About 170 miles long and up to 36 miles wide it is just north of the great crater Copernicus. The higher peaks are up to 1½ miles high. Observe these fine mountains 2 days after first quarter. They make a grand sight in any size scope.

Gay Lussac (16 miles) is a low walled crater in the Carpathians. The rim is only about ½ mile above the flat floor.

Draper (5 miles) is a deep little crater just north of the Carpathians. At 1 mile deep it makes a nice pairing with an unnamed crater that is very similar to itself and located 5 miles to its south.

Pytheas (12 miles) is a prominent crater that is about 1½ mile deep. The floor is rough and rather small.

D. Spain, *The Six-Inch Lunar Atlas: A Pocket Field Guide*,
DOI 10.1007/978-0-387-87610-8_46, 

Upright, North up - naked eye, binoculars, spotting scopes
Pytheas
Montes
Draper
Carpatus
Gay Lussac

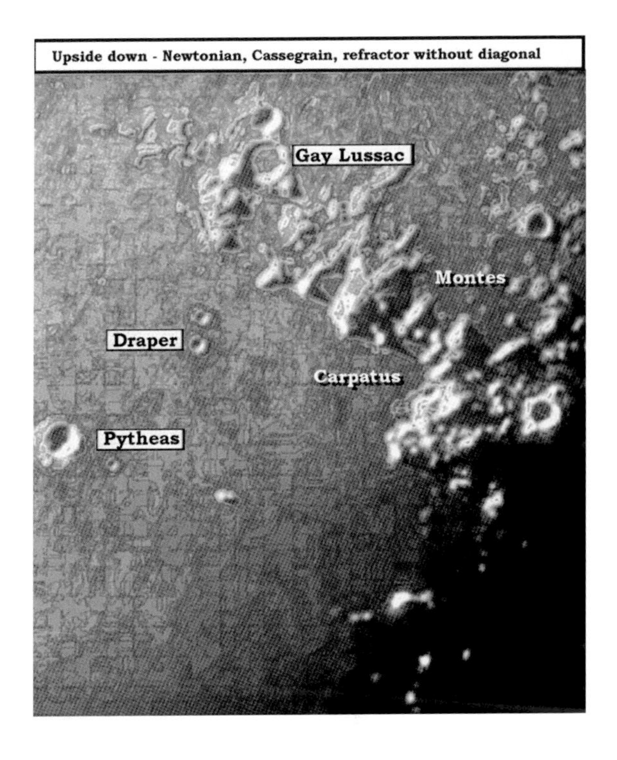
Upside down - Newtonian, Cassegrain, refractor without diagonal
Gay Lussac
Montes
Draper
Carpatus
Pytheas

Upright, left/right reversed - refractors/catadioptrics with diagonal
Pytheas
Montes
Draper
Carpatus
Gay Lussac

## CHART 45

# COPERNICUS

Copernicus (56 miles) is one of the greatest showpiece formations on the lunar surface. It is like Tycho in age, probably a little over 1 billion years, making it a youngster when compared to most other craters. It makes its appearance 2 days after first quarter. Careful examination of its rim shows that its shape is almost hexagonal. From the bottom of its flat floor to the top of its rim the heavily terraced walls rise over two miles. There are three main central peaks, the highest reaching up ¾ of a mile. The top of the rim is about ½ mile above the surrounding plain. Spreading outward from the base of the walls are ridges in a radial pattern. This one is not to be missed. Copernicus is to lunar observers what the Andromeda Galaxy is to deep sky observers. Observe it with any and all telescopes. It will never disappoint you.

Fauth (7 miles) is conjoined with a 6 mile crater directly below it. Fauth is a little over a mile deep and its little brother is just under a mile. Together they look like a keyhole.

Reinhold (29 miles) is a sharp rimmed and prominent crater. Its flat floor has terraced walls that rise almost 2 miles. My 6 inch refractor reveals a tiny off center peak.

D. Spain, *The Six-Inch Lunar Atlas: A Pocket Field Guide*,
DOI 10.1007/978-0-387-87610-8_47, 

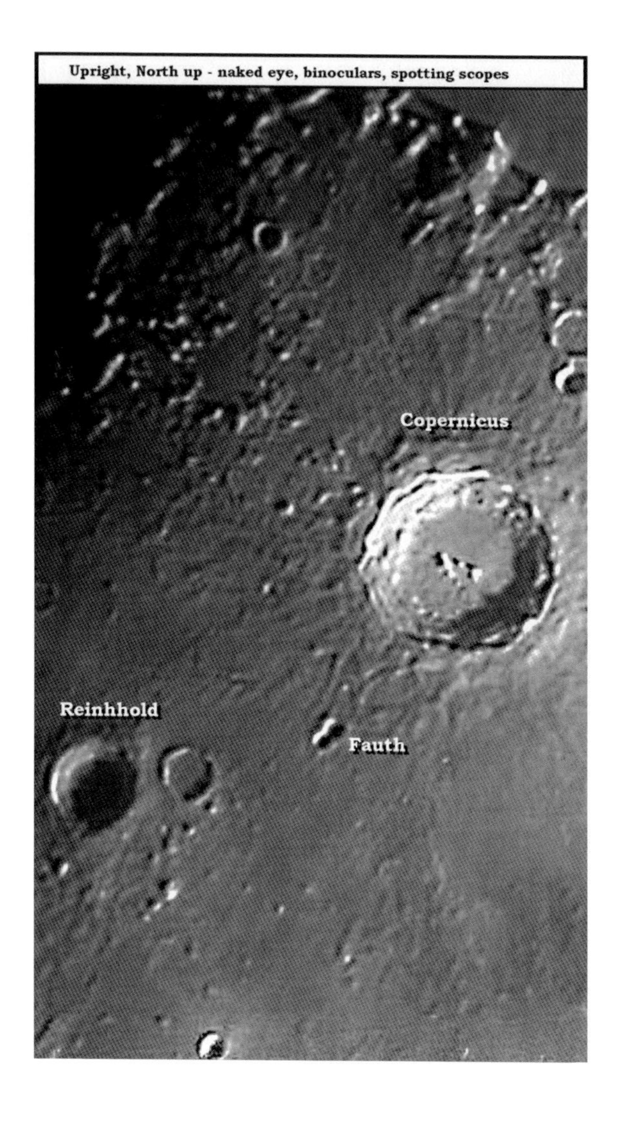
Upright, North up - naked eye, binoculars, spotting scopes
Copernicus
Reinhhold
Fauth

Upside down - Newtonian, Cassegrain, refractor without diagonal
Fauth
Reinhold
Copernicus

Upright, left/right reversed - refractors/catadioptrics with diagonal
Copernicus
Reinhold
Fauth

## CHART 46

# BULLIALDUS

Bullialdus (37 miles) has the appearance of a miniature Copernicus. Almost 2 miles deep it has a central mountain with several peaks. Its walls are terraced and it has outside ridges in a similar radial pattern. While Copernicus demands your attention 2 days after first quarter, take the time to travel southward to this prominent formation.

Lubiniezky (27 miles) is a flooded crater to Bullialdus' northwest. The walls are very low, not reaching ½ mile high. The flat floor is devoid of any detail in my 6 inch refractor.

Konig (14 miles) is a prominent little crater southwest of Bullialdus. Sharp rimmed and deep, the walls are almost 1½ miles high. This is a charming little crater.

Explore this area at your leisure. Note the pair of unnamed craters below Bullialdus. There are several little hills and craters that are waiting for you to discover.

D. Spain, *The Six-Inch Lunar Atlas: A Pocket Field Guide*,
DOI 10.1007/978-0-387-87610-8_48, 

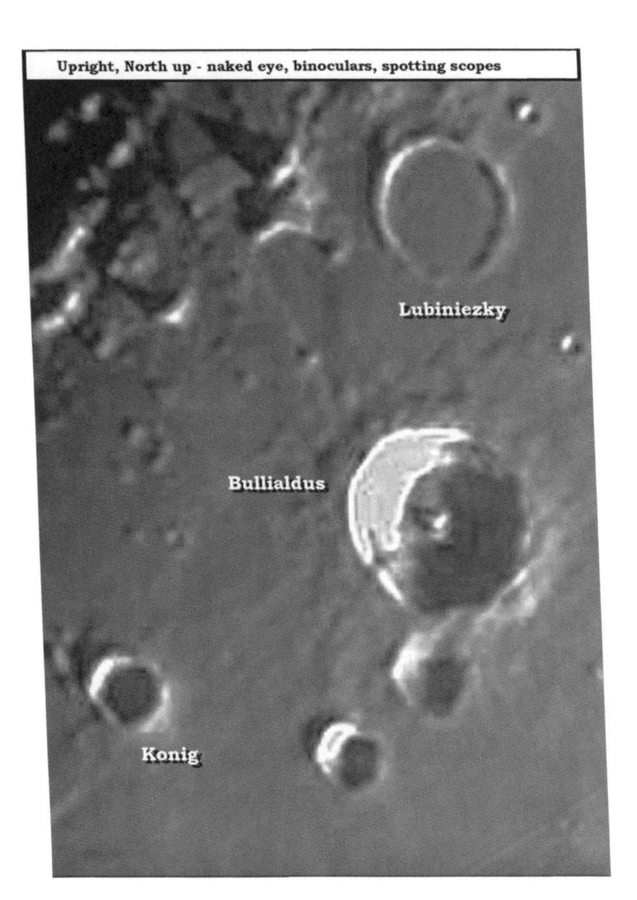
Upright, North up - naked eye, binoculars, spotting scopes
Lubiniezky
Bullialdus
Konig

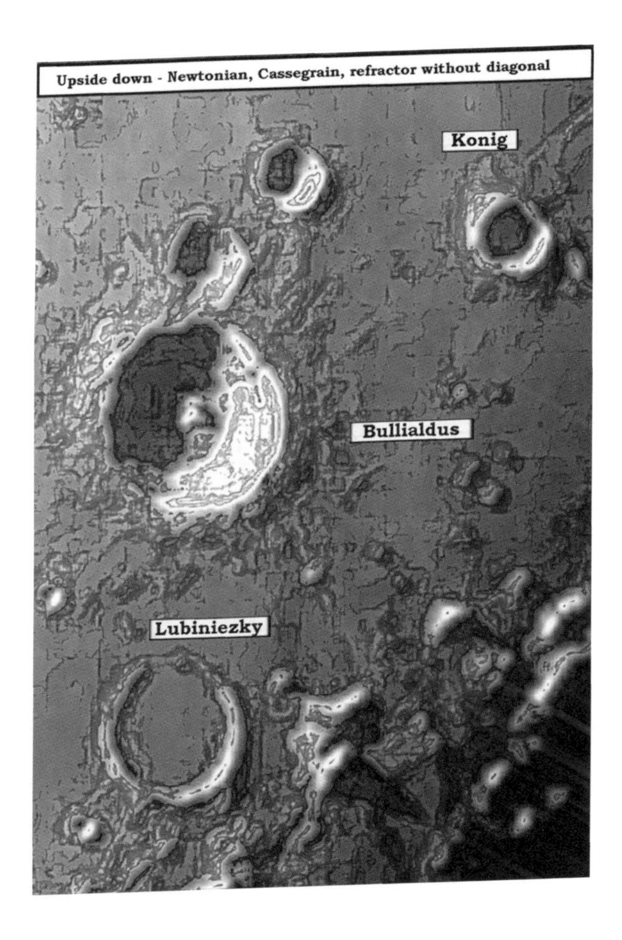
Upside down - Newtonian, Cassegrain, refractor without diagonal
Konig
Bullialdus
Lubiniezky

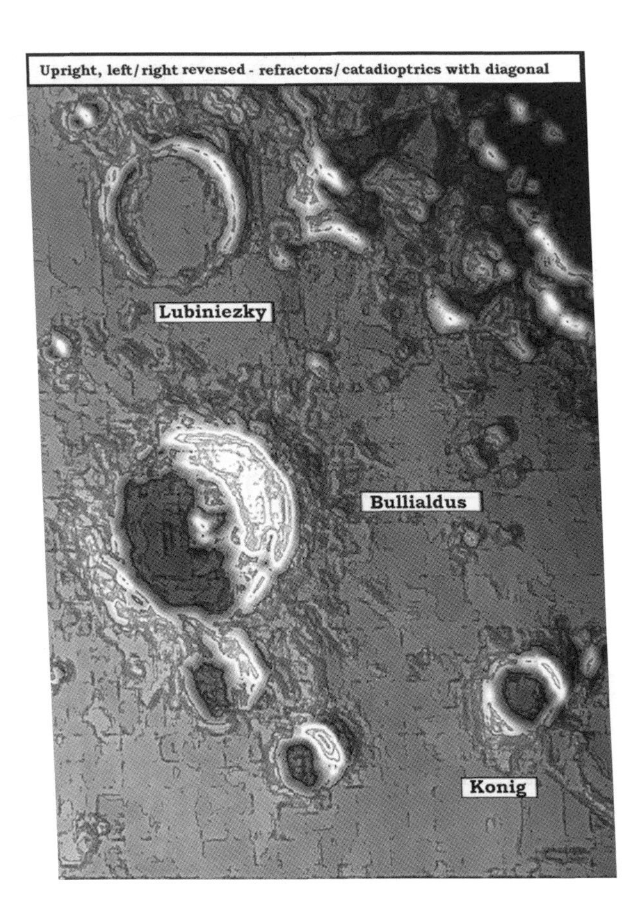
Upright, left/right reversed - refractors/catadioptrics with diagonal
Lubiniezky
Bullialdus
Konig

## CHART 47

# MONTES RIPHAEUS

Montes Riphaeus or the Riphaean Mountains is an isolated range on the Oceanus Procellarum, the Ocean of Storms. Running in a north-south direction these low peaks are not particularly high, the tallest is just short of being 1 mile above the plain. Still, when the sun is rising here 2 days after first quarter be sure to explore this little range.

Euclides (7 miles) is a sharp and very prominent crater. Its rim is a little over ½ mile above its floor. Located due west of the Riphaeus, you will have no problem locating it.

Darney (9 miles) is the other prominent crater in this region. It is quite deep for its diameter. The sharp rim is 1½ mile above the floor.

If you can tear yourself away from Copernicus and Bullialdus, this region will reward you with pleasant views of its lonely hills, peaks and craters.

D. Spain, *The Six-Inch Lunar Atlas: A Pocket Field Guide*,
DOI 10.1007/978-0-387-87610-8_49, 

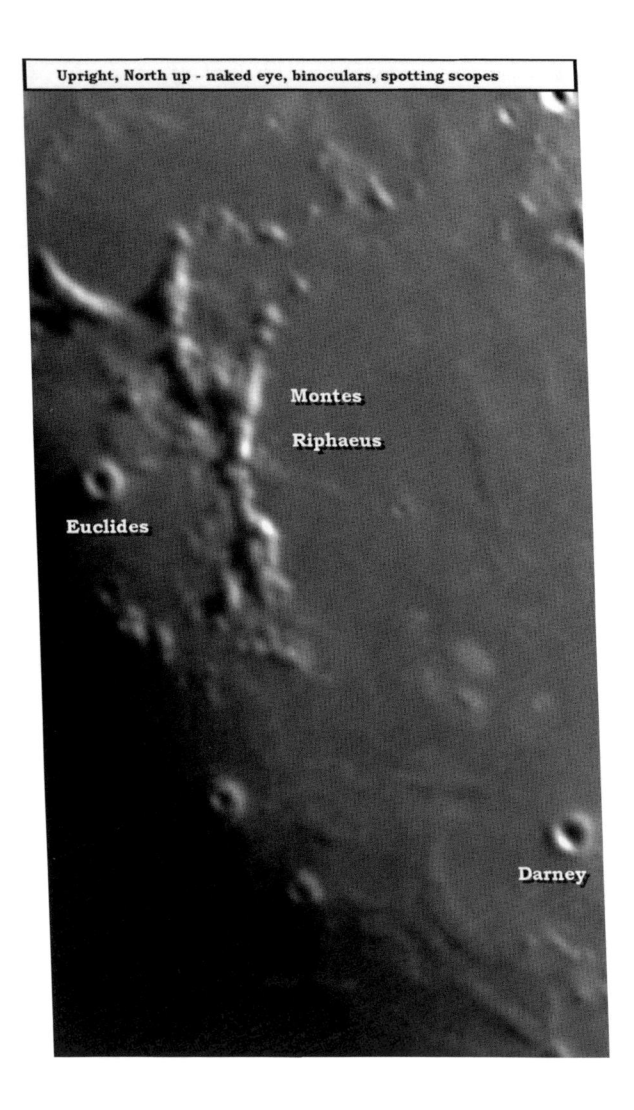
Upright, North up - naked eye, binoculars, spotting scopes
Montes
Riphaeus
Euclides
Darney

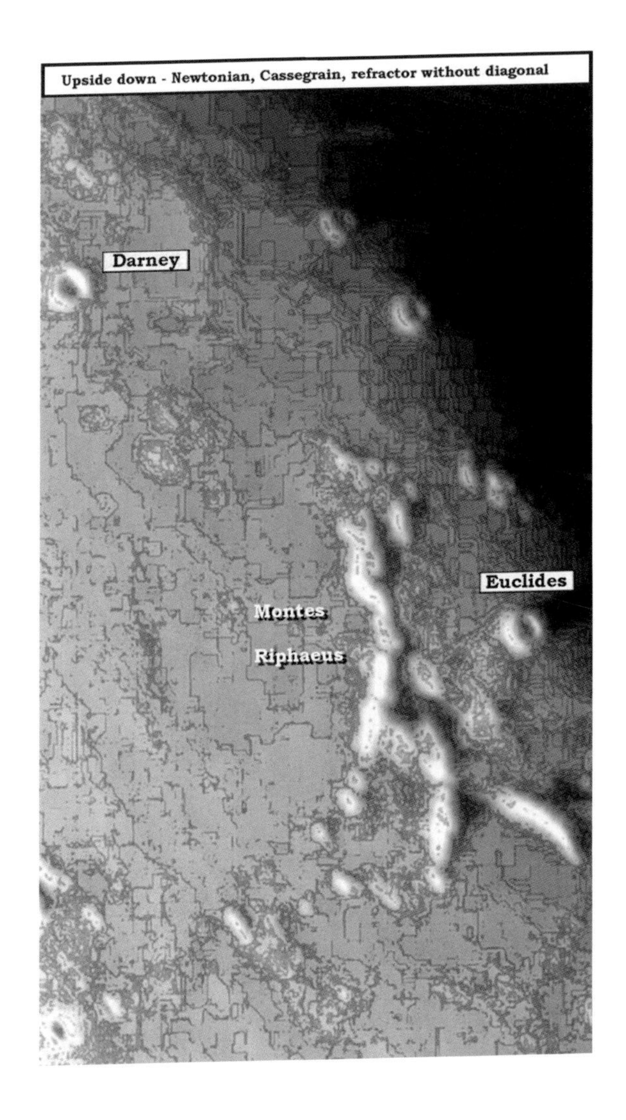
Upside down - Newtonian, Cassegrain, refractor without diagonal
Darney
Euclides
Montes
Riphaeus

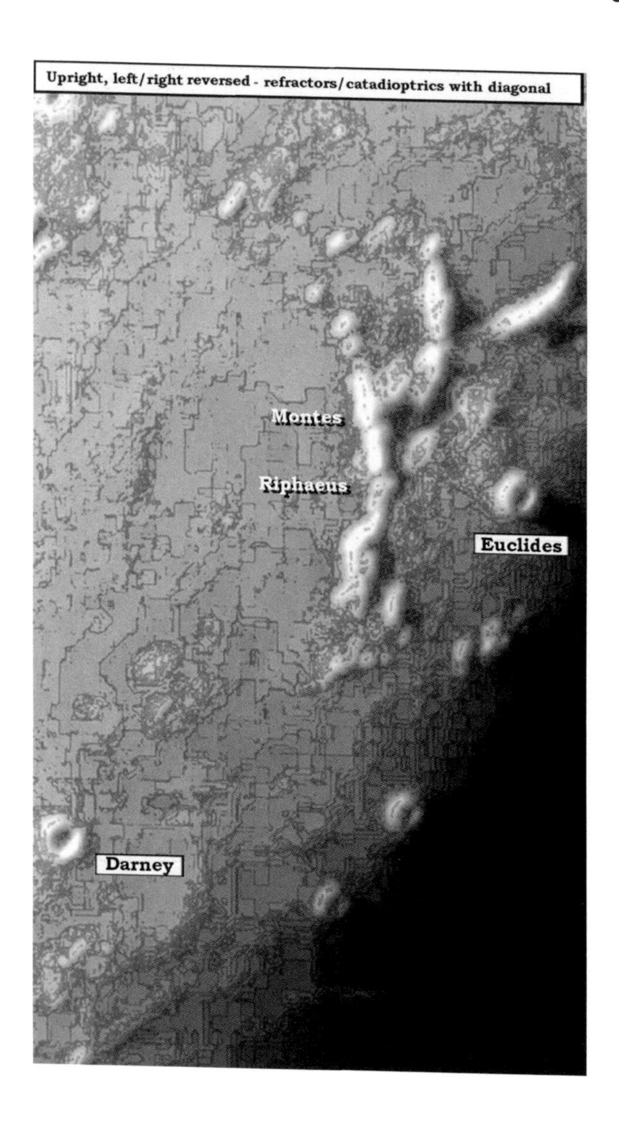
Upright, left/right reversed - refractors/catadioptrics with diagonal
Montes
Riphaeus
Euclides
Darney

## CHART 48

# J. HERSCHEL

J. Herschel, (95 miles) is a great ruined formation near the top of the Moon. The walls are blocky and in some places are open to the outside surroundings. My estimation of the height of the wall is not much above 1 mile. The floor is rough and uneven. There are two easily seen craters east of central on the floor. The larger is 7 miles across and the smaller 5 miles. This old giant doesn't get many visitors, so if you can; take a journey to it 3 days after first quarter.

Horrebow (15 miles) is on the southwest rim of J. Herschel. The walls are about 1½ mile high and the floor looks bumpy through my 6 inch refractor.

D. Spain, *The Six-Inch Lunar Atlas: A Pocket Field Guide*,
DOI 10.1007/978-0-387-87610-8_50, © Springer Science+Business Media, LLC 2009

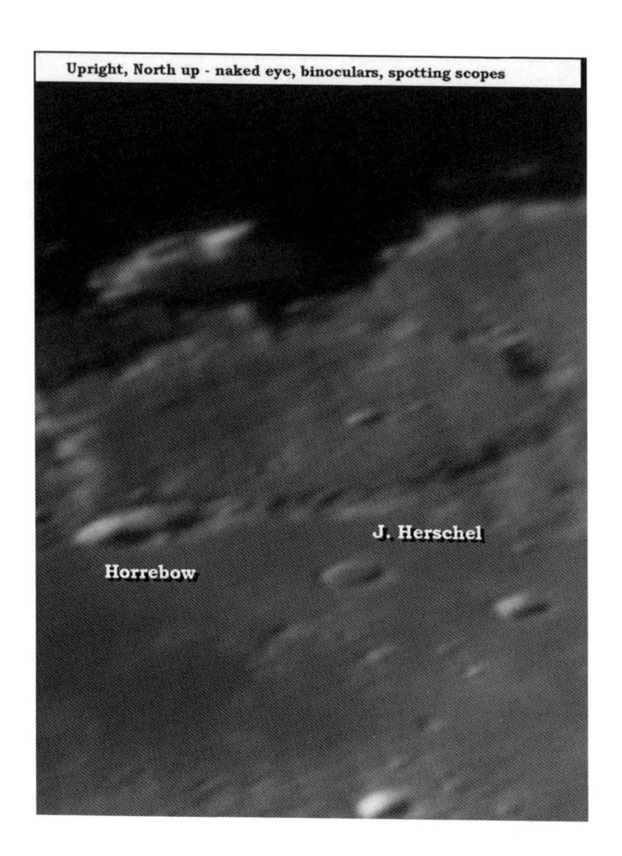
Upright, North up - naked eye, binoculars, spotting scopes
J. Herschel
Horrebow

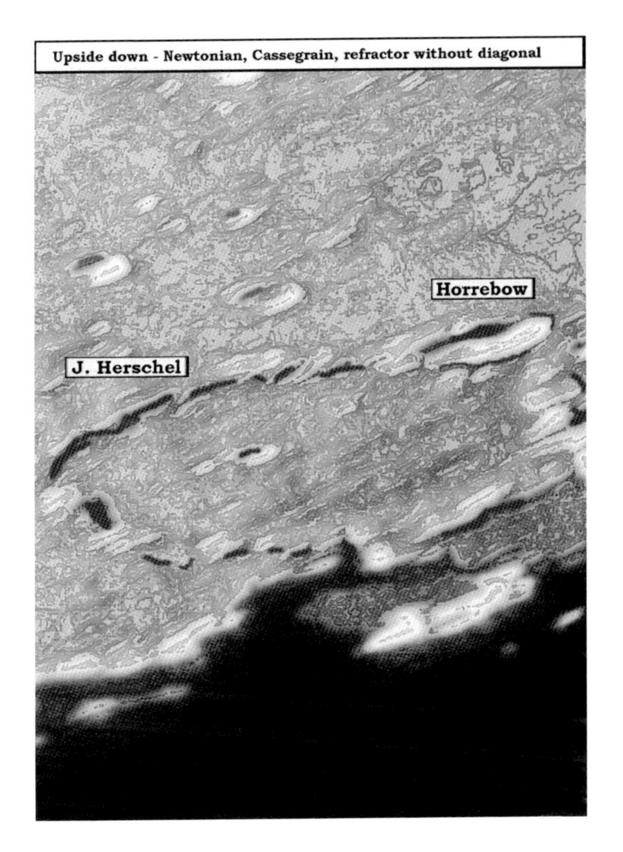
Upside down - Newtonian, Cassegrain, refractor without diagonal
Horrebow
J. Herschel

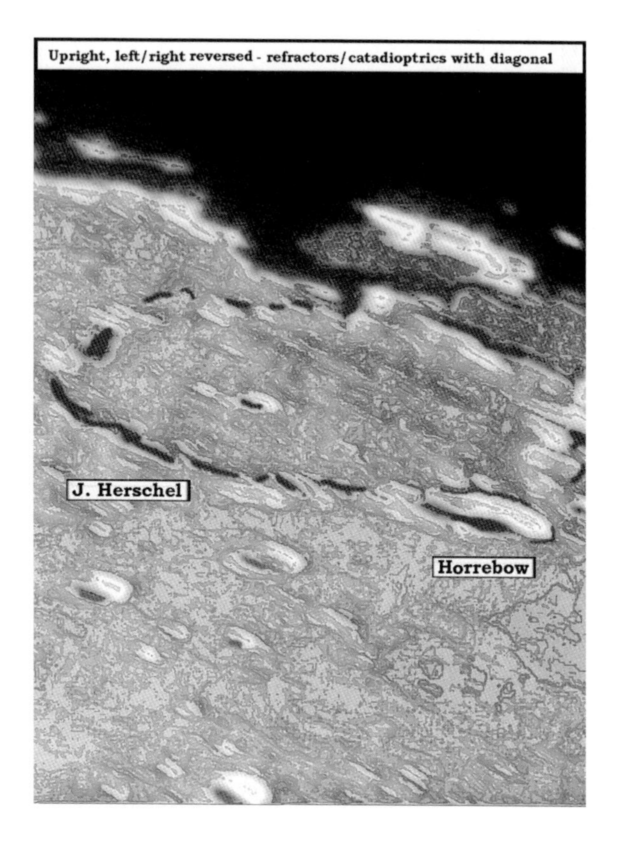
Upright, left/right reversed - refractors/catadioptrics with diagonal
J. Herschel
Horrebow

# CHART 49

## SINUS IRIDUM

Sinus Iridum is a beautiful bay off the northwestern boundary of Mare Imbrium, the Sea of Rains. Sinus Iridum means the Bay of Rainbows and while you will never see any rainbows here, the bay is fascinating. It was formed by a colossal impact billions of years ago. The distance between the two promontories is over 140 miles. Imagine how imposing this formation would have looked if the southern wall were still intact. When observing this bay 3 days after first quarter, the northern wall will be illuminated by the rising sun while part of the bay will still be in darkness. Notice the long wrinkle ridges. They remind me of waves of water flowing into the bay. With just a little imagination I can see these ridges as great breakers rolling into the bay and splashing onto the shore. You will come back here time and time again to surf on this wonderful bay.

Promontory Heraclides is the cape that guards the western edge of the Bay. Standing 1 mile above the bay this towering peak it is a grand sight.

Promontory Laplace guards the eastern entrance to the bay. It reaches 1½ mile above the bay and casts a very long triangular shadow at sunrise.

This is one of the most beautiful areas on the Moon at sunrise. It is an absolute must show to any and all guests at your scope.

D. Spain, *The Six-Inch Lunar Atlas: A Pocket Field Guide*,
DOI 10.1007/978-0-387-87610-8_51, 

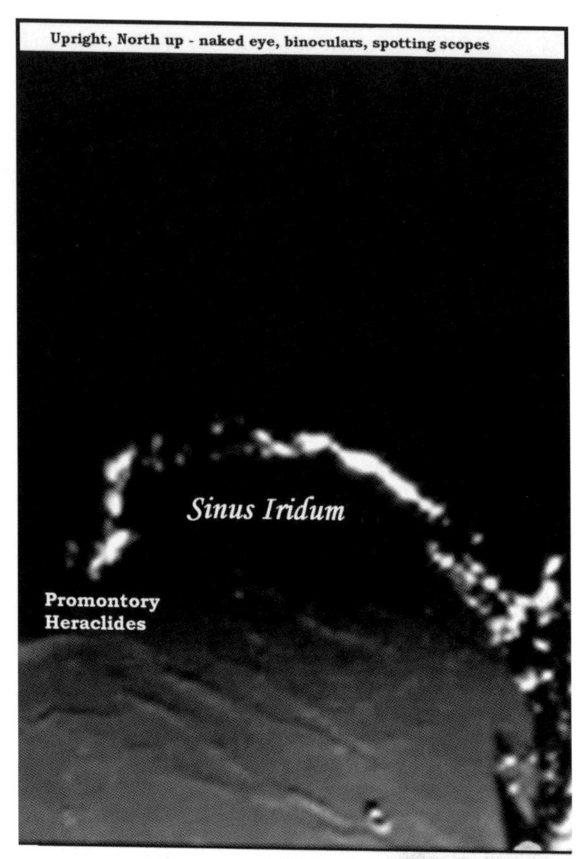
Upright, North up - naked eye, binoculars, spotting scopes
Sinus Iridum
Promontory
Heraclides
Promontory
Laplace

**Upside down - Newtonian, Cassegrain, refractor without diagonal**

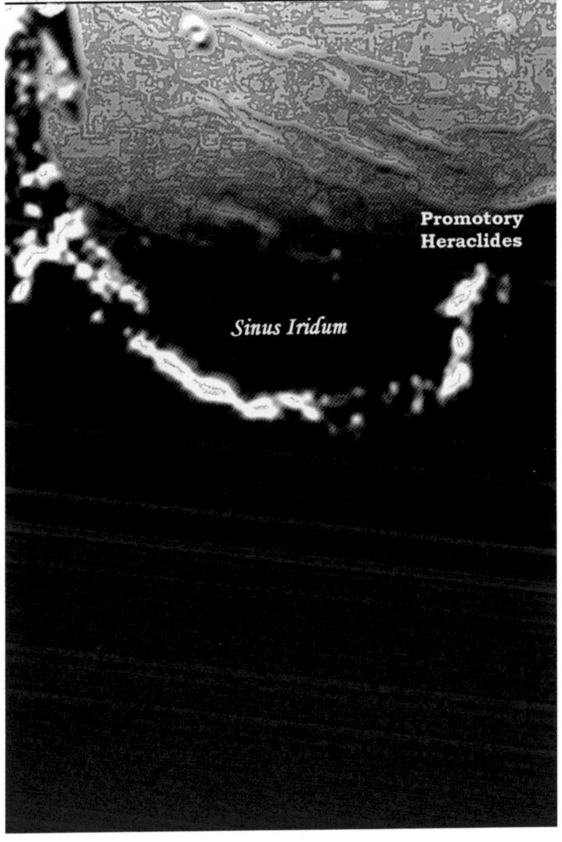

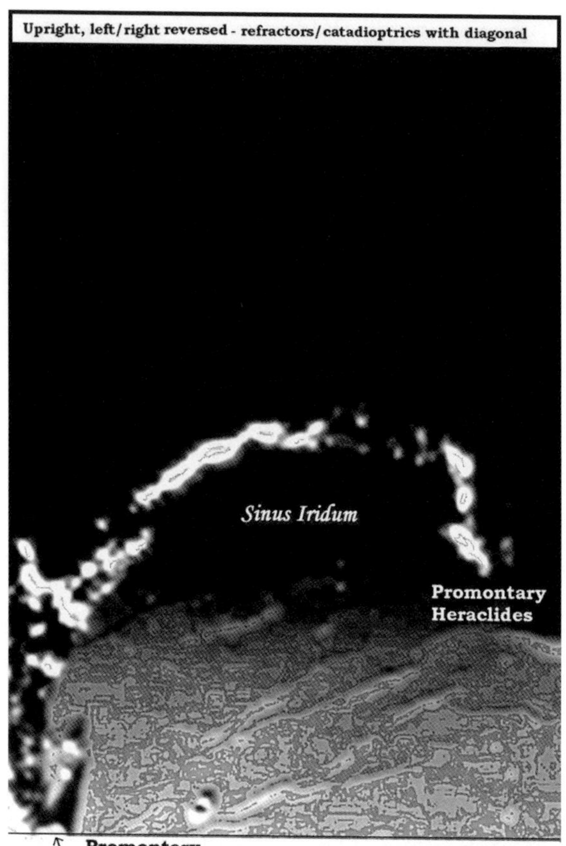
Upright, left/right reversed - refractors/catadioptrics with diagonal
Sinus Iridum
Promontary
Heraclides
Promontary
Laplace

## CHART 50

# GASSENDI

Gassendi (68 miles) is a fine old crater that has seen better times. It is my personal favorite formation on the lunar surface. The walls are just over 1 mile high, but they are very low on the southern rim, where they are just a few hundred feet high. The large flat floor is full of detail. Three central peaks dominate the floor. The highest is about ¾ mile. The peaks cast long shadows across the floor 3 days after first quarter. The floor is much higher in the north and is bumpy and rough. The walls encompass lots of hills and ridges and the floor has a great system of rills. Some of the rills east of the central peaks can be glimpsed in my 4 inch refractor, but it takes the 6 inch to bring out the fainter and delicate rills.

There is a great 20 mile crater on Gassendi's north rim that is a fine crater by itself. It is about 2 miles deep and has terraced walls around a central peak.

The first time I saw Gassendi it reminded me of a great medieval castle. The rim was the great walls of stone and the central mountains were the castle's Keep. The low walls to the south were where the drawbridge keeps the enemy out.

Gassendi is one of the most lovely and fascinating formations on the Moon. It begs you to come and explore its kingdom. Once you find it you will come back time and time again.

D. Spain, *The Six-Inch Lunar Atlas: A Pocket Field Guide*,
DOI 10.1007/978-0-387-87610-8_52, © Springer Science+Business Media, LLC 2009

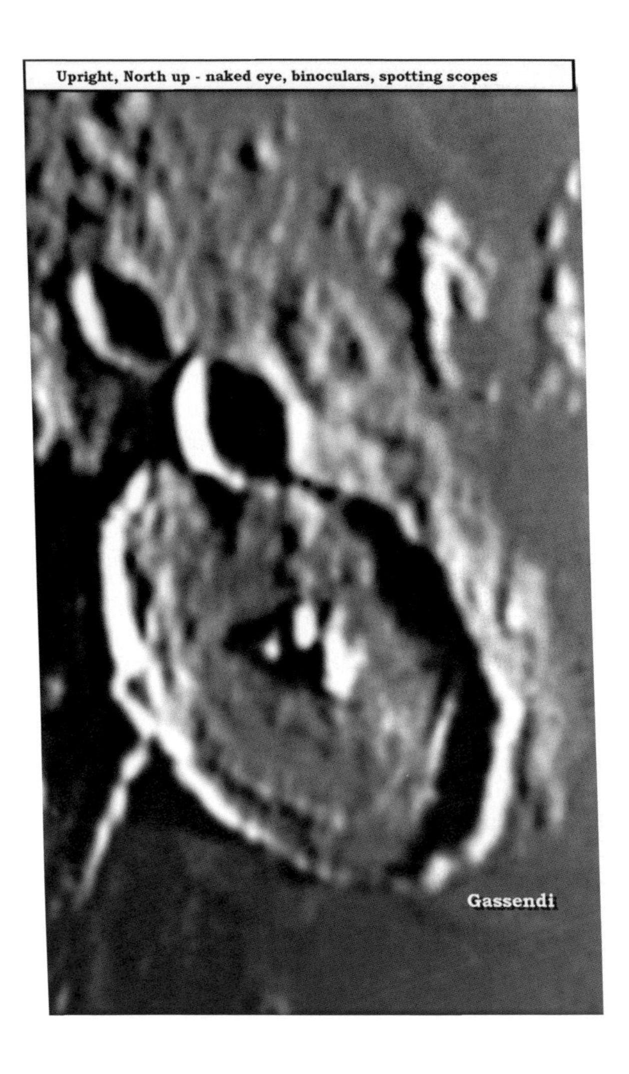
Upright, North up - naked eye, binoculars, spotting scopes
Gassendi

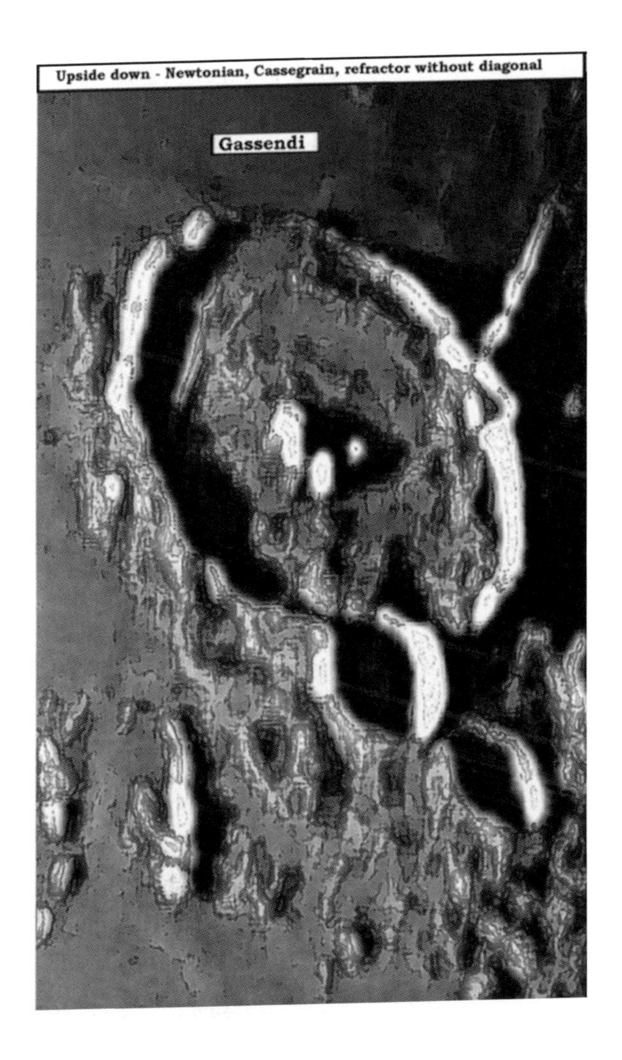
Upside down - Newtonian, Cassegrain, refractor without diagonal
Gassendi

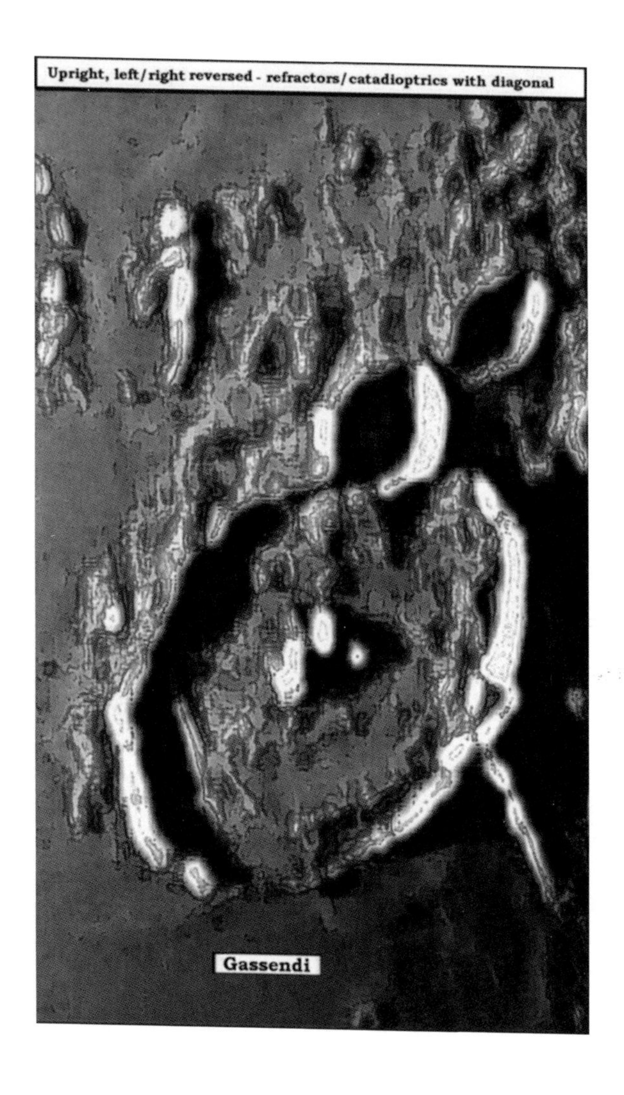
Upright, left/right reversed - refractors/catadioptrics with diagonal
Gassendi

## CHART 51

# SCHILLER

Schiller (108 by 43 miles) is a greatly elongated formation. The walls are over 2 miles above the floor and terraced along the southern slopes. The floor is flat except in the north. The north floor has two sizable separate mountain masses. This formation may be from the fusion of two craters or from a very low angle impact by a small asteroid. Examine Schiller on the third day after first quarter and see which you think it is.

Bayer (29 miles) is just off the southeast rim of Schiller. Its walls rise to over 1 mile above the flat floor. There is a nice little 4 mile crater on the south floor, easily visible with my 2.4 inch refractor.

Rost (30 miles) is a flat floor crater with walls that reach a little over a mile to the rim. I see no detail on the floor, even in my 6 inch refractor.

Segner (40 miles) is a shallow crater for its diameter. Its walls are only ¾ mile high. There is a little 4 mile crater near the center of the rough floor.

Zucchius (39 miles) is much deeper that its cousin Segner to its northeast. Its terraced walls are almost 2 miles about the floor. There is a nice central peak on the rough floor.

D. Spain, *The Six-Inch Lunar Atlas: A Pocket Field Guide*,
DOI 10.1007/978-0-387-87610-8_53, © Springer Science+Business Media, LLC 2009

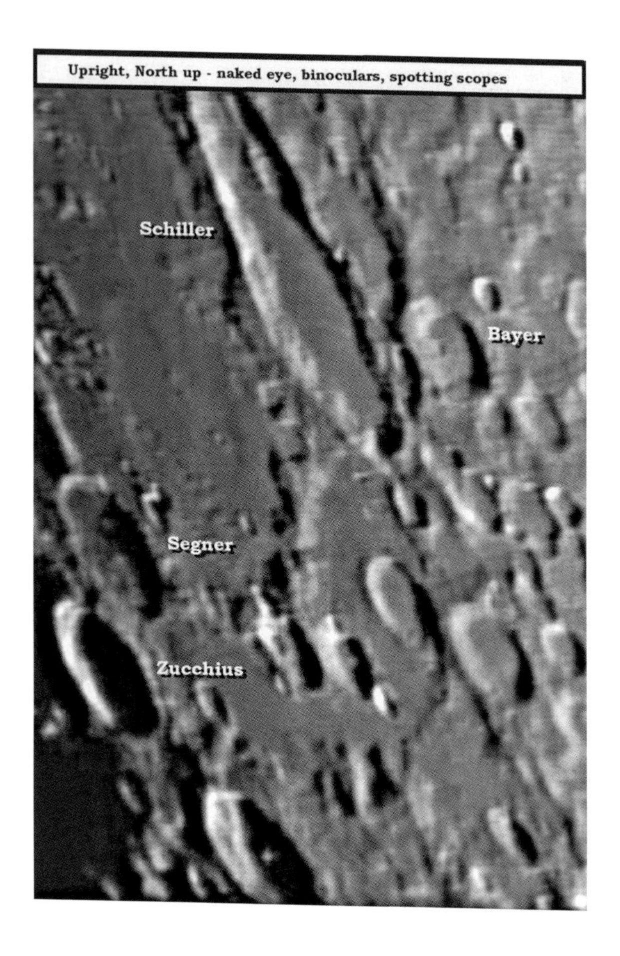
Upright, North up - naked eye, binoculars, spotting scopes
Schiller
Bayer
Segner
Zucchius

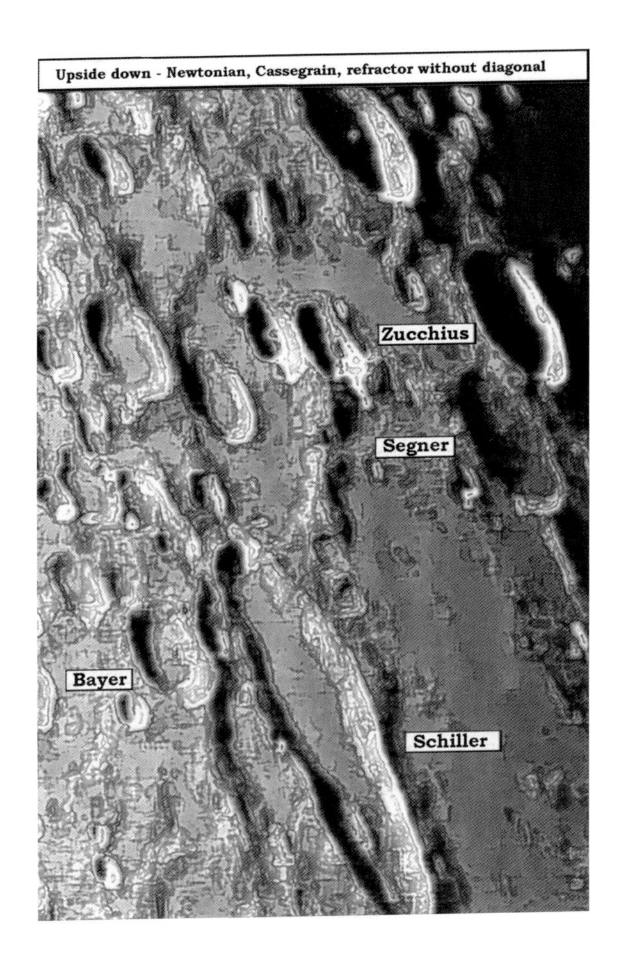
Upside down - Newtonian, Cassegrain, refractor without diagonal
Zucchius
Segner
Bayer
Schiller

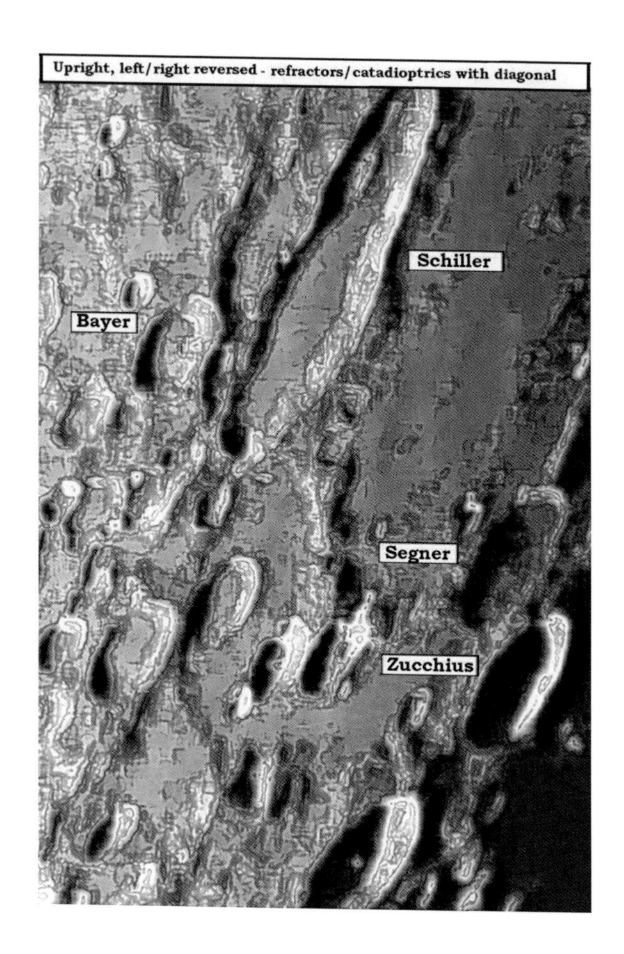
Upright, left/right reversed - refractors/catadioptrics with diagonal
Schiller
Bayer
Segner
Zucchius

## CHART 52

# SCHICKARD

Schickard (137 miles) is another of the lunar giant formations. The walls are broad and rather steep. I would guess their height at 1-1½ miles. The immense flat floor is darker in the north and there is another dark patch in the southeast. There are scores of craters and craterlets on the floor and the walls of this gentle monster. When viewing it 5 days after first quarter under the lunar sunrise even my 2.4 inch refractor will show great differences in topology. This is one of the formations that are must viewing for your friends and quests.

Nasmyth (47 miles) pales in comparison to its northern neighbor Schickard. Still this is a big formation that is worth viewing. There is a small crater near the center of the floor and an 8 mile crater on its northern rim. Its southern wall is destroyed by Phocylides.

Phocylides (70 miles) is a great formation below Nasmyth. Its broad walls surround a flat floor studded with craters and craterlets. Careful inspection will show the floor is somewhat darker in the south.

Wargentin (51 miles) is in a unique class of craters. It is the largest of craters filled to the rim with lava. It is like a great mesa ½ mile above the surrounding region. My 4 inch refractor will reveal a Y shaped wrinkle ridge on the surface under the early morning lunar sun. This is an exceptional formation that you must share with everyone.

D. Spain, *The Six-Inch Lunar Atlas: A Pocket Field Guide*,
DOI 10.1007/978-0-387-87610-8_54, 

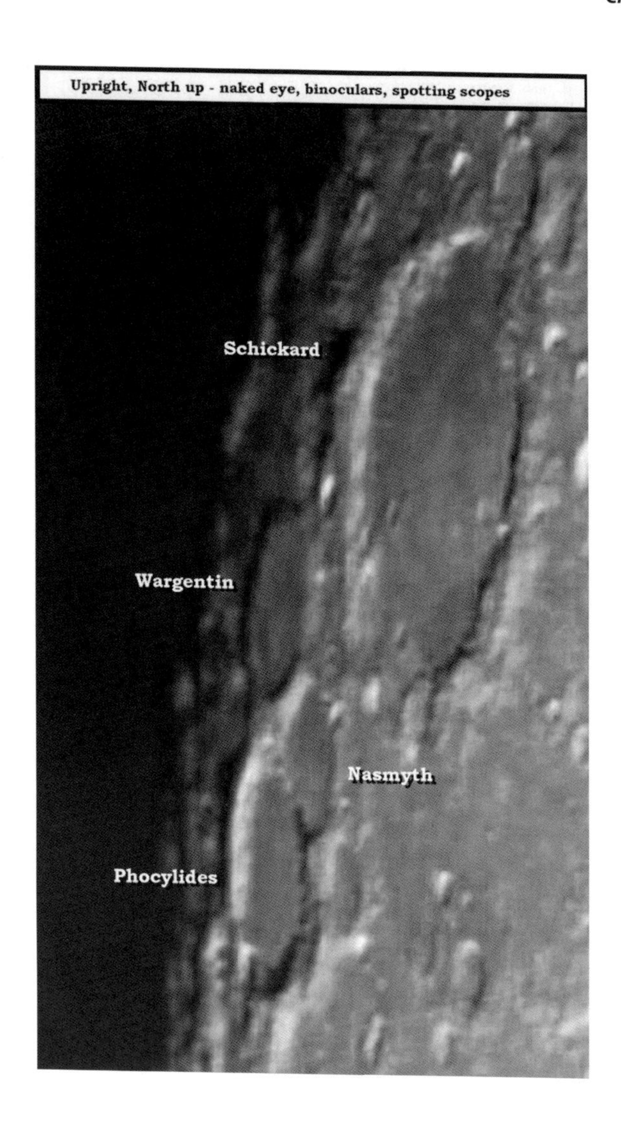
Upright, North up - naked eye, binoculars, spotting scopes
Schickard
Wargentin
Nasmyth
Phocylides

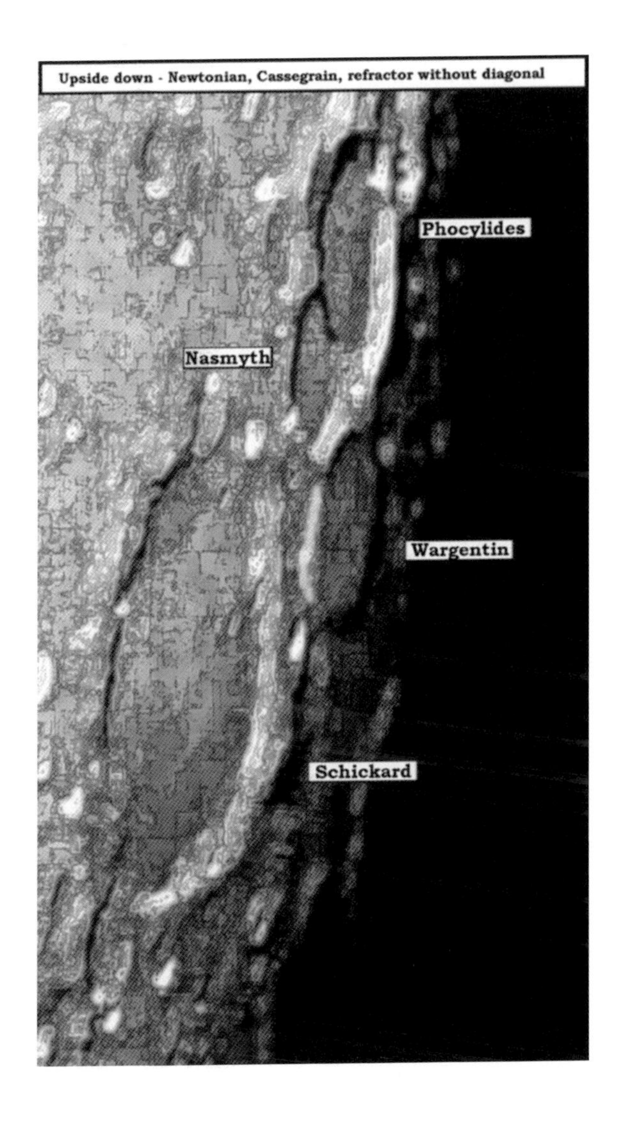
Upside down - Newtonian, Cassegrain, refractor without diagonal
Phocylides
Nasmyth
Wargentin
Schickard

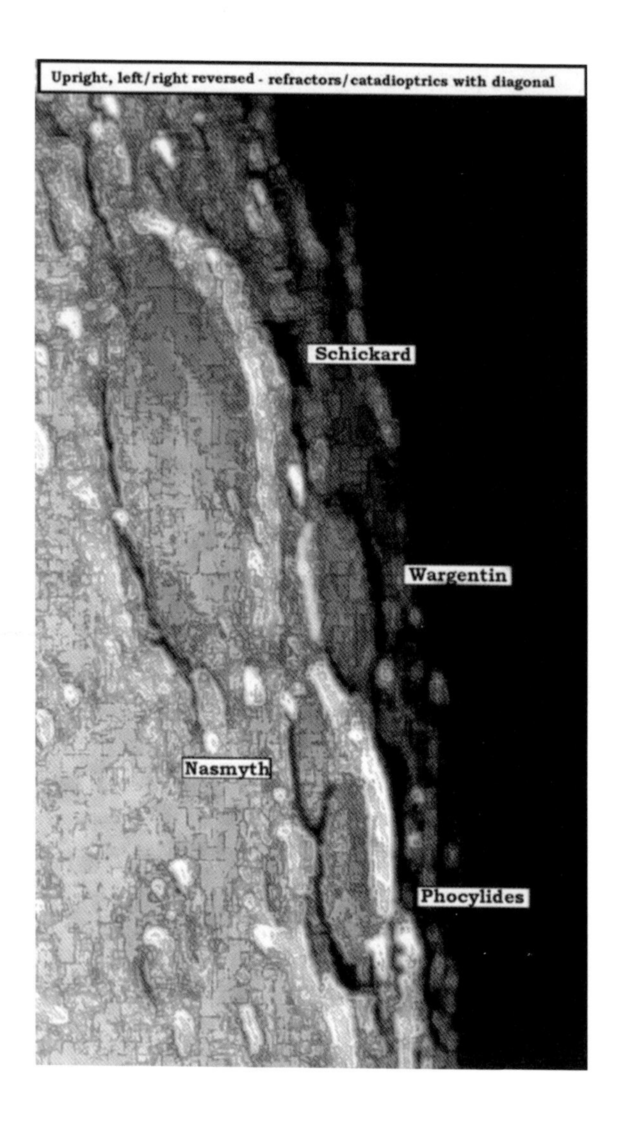
Upright, left/right reversed - refractors/catadioptrics with diagonal
Schickard
Wargentin
Nasmyth
Phocylides

## CHART 53

# PYTHAGORAS

Pythagoras (78 miles) is a large crater tucked away near the top of the Moon. Almost 3 miles deep, the terraced walls enclose a large flat floor. At 5 days after first quarter it is an imposing formation. Notice the double central mountain. Because it is near the edge of the Moon it is like looking at it almost in profile and unfortunately, it is usually ignored by the casual lunar observer.

Babbage (87 miles) is an irregularly shaped and worn down formation. The walls are low and the floor rough. There is about a 20 mile crater on the floor and another 8 mile crater close to it.

Oenopides (40 miles) is on the southern boarder of Babbage. The walls are low and the floor is pretty smooth.

Harpalus (24 miles) is a deep, well formed crater with terraced walls over 1½ miles high. The floor is flat and there is a central mountain mass.

D. Spain, *The Six-Inch Lunar Atlas: A Pocket Field Guide*,
DOI 10.1007/978-0-387-87610-8_55, 

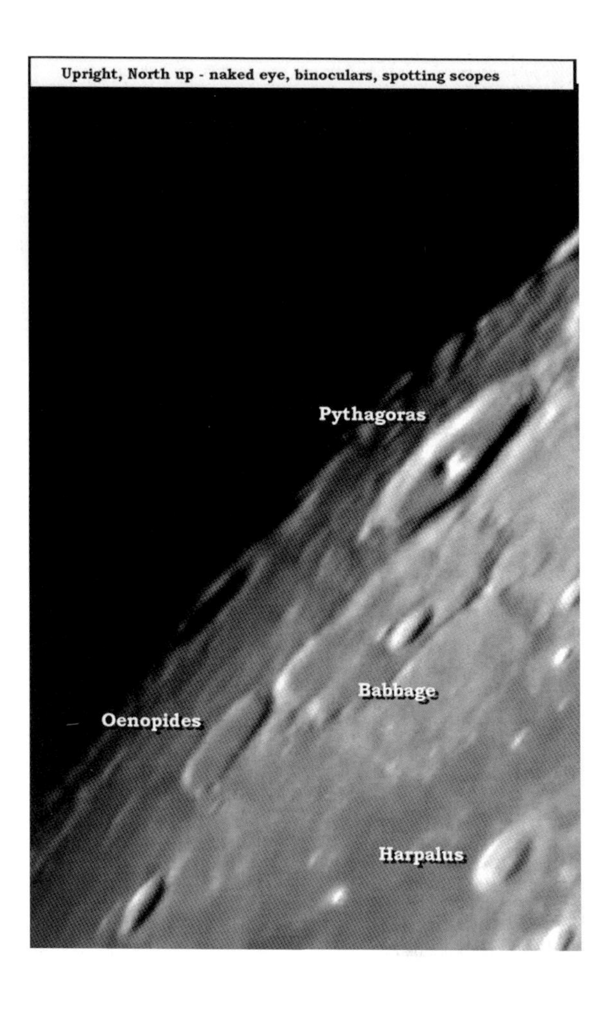
Upright, North up - naked eye, binoculars, spotting scopes
Pythagoras
Babbage
Oenopides
Harpalus

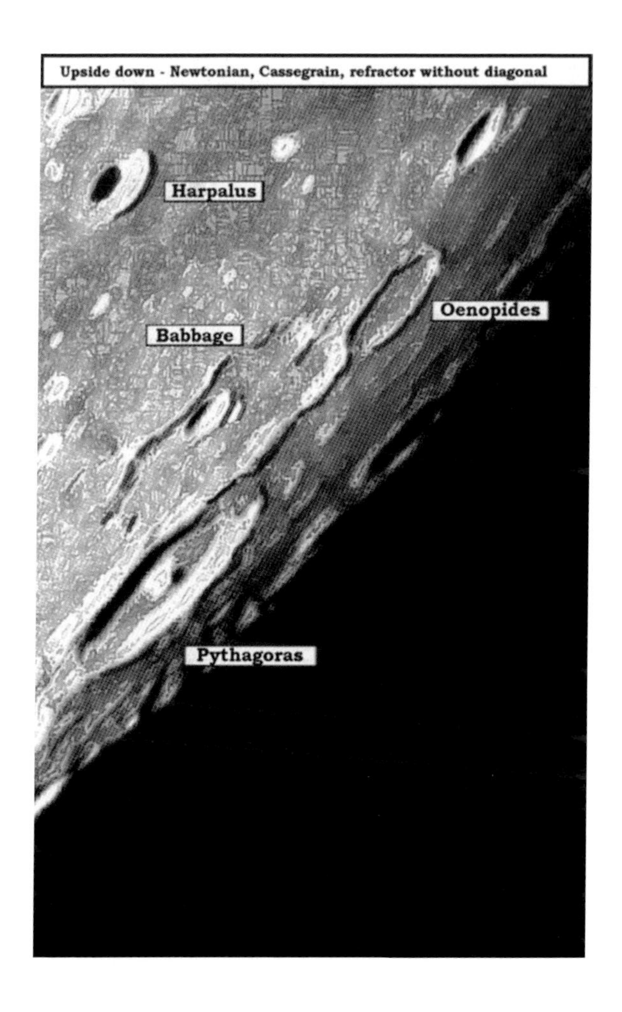
Upside down - Newtonian, Cassegrain, refractor without diagonal
Harpalus
Oenopides
Babbage
Pythagoras

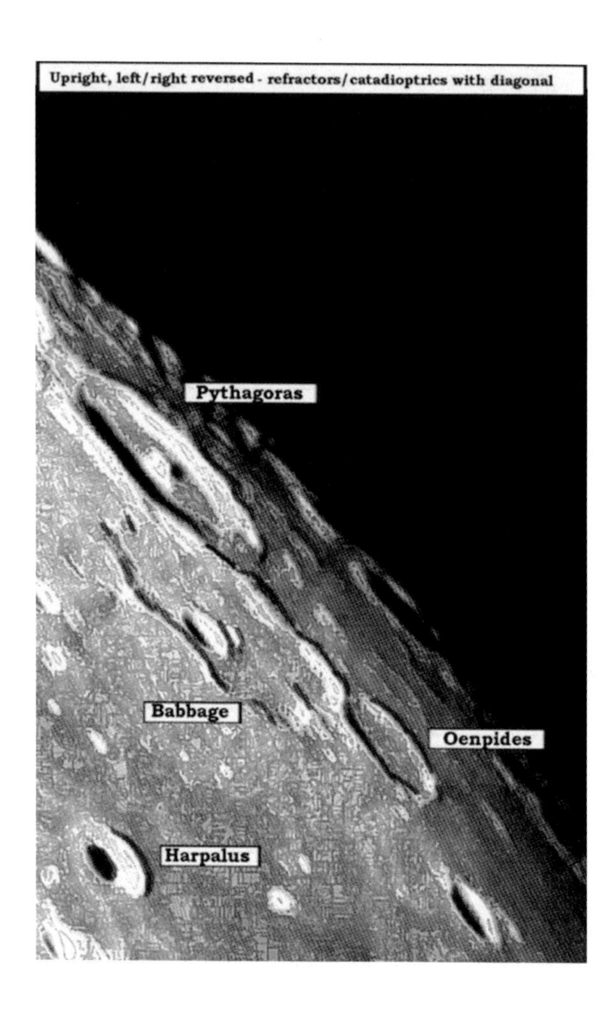
Upright, left/right reversed - refractors/catadioptrics with diagonal
Pythagoras
Babbage
Oenpides
Harpalus

## CHART 54

# ARISTARCHUS

Aristarchus (25 miles) is a rather small, but deep crater. Its terraced walls are about 1¾ miles above the flat floor. There is small central mountain in this young crater. As the sun rises on Aristarchus 4 days after first quarter you will notice that the interior walls are very white. In fact, this is one of the brightest and whitest formations on the Moon. The keen eyed observer with a 2.4 inch refractor may notice faint bands on the western wall. These bands run vertically and are easy in any 4 inch and larger scope. As the sun gets higher they become invisible.

Herodotus (21 miles) is a much older crater than Aristarchus. Its walls are low and are less than 1 mile high. The floor is dark with little detail visible even in my 6 inch refractor.

Schroter's Valley, officially named as Vallis Schroteri is the largest rill on the Moon. It winds for over 100 miles and is visible in small scopes. At its beginning, commonly called the Cobra Head, it is nearly 4 miles wide, but eventually narrows to about 1/3 mile. It is a little over 1/2 mile deep but is much lower at its end. This is a prize formation to show to your guests.

Prinz (29 miles) is a crater missing its southern rim. Its low walls are just over 1/2 mile high and the floor is flooded with lava.

The Harbinger Mountains are a small group of peaks northeast of Prinz. Not exceeding 1½ miles in height they are still an attractive sight at lunar sunrise.

Krieger (13 miles) is another flooded crater with low walls. It has a 5 mile crater on its southeastern rim.

Angstrom (6 miles) and Nielsen (6 miles) are both deep, sharp rimmed craters.

Nielsen is located on the southern end of a long wrinkle ridge.

D. Spain, *The Six-Inch Lunar Atlas: A Pocket Field Guide*,
DOI 10.1007/978-0-387-87610-8_56, 

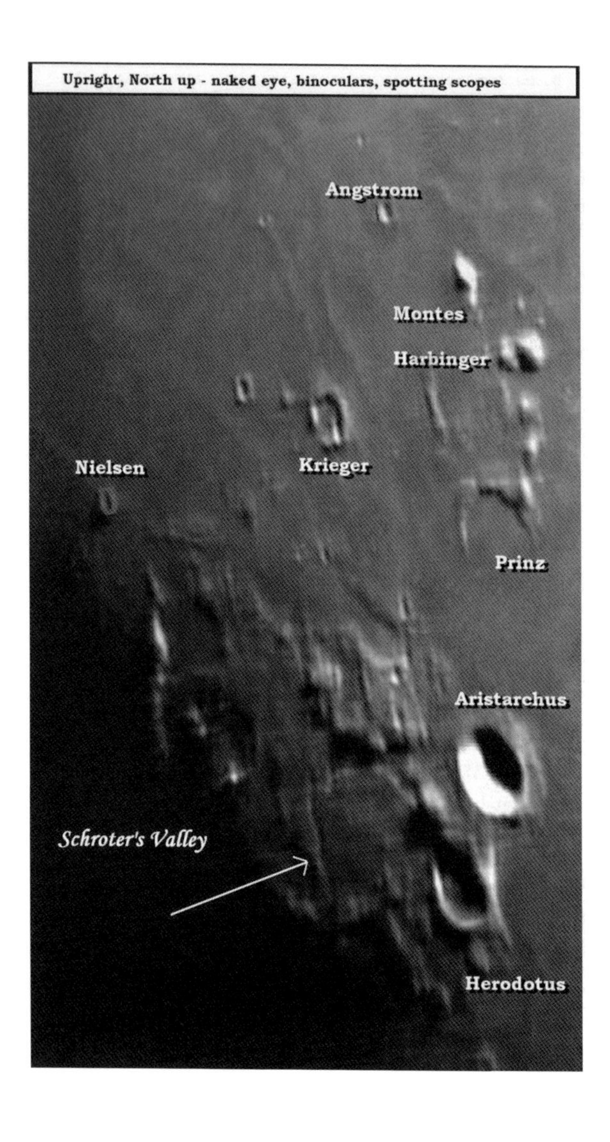
Upright, North up - naked eye, binoculars, spotting scopes
Angstrom
Montes
Harbinger
Nielsen
Krieger
Prinz
Aristarchus
Schroter's Valley
Herodotus

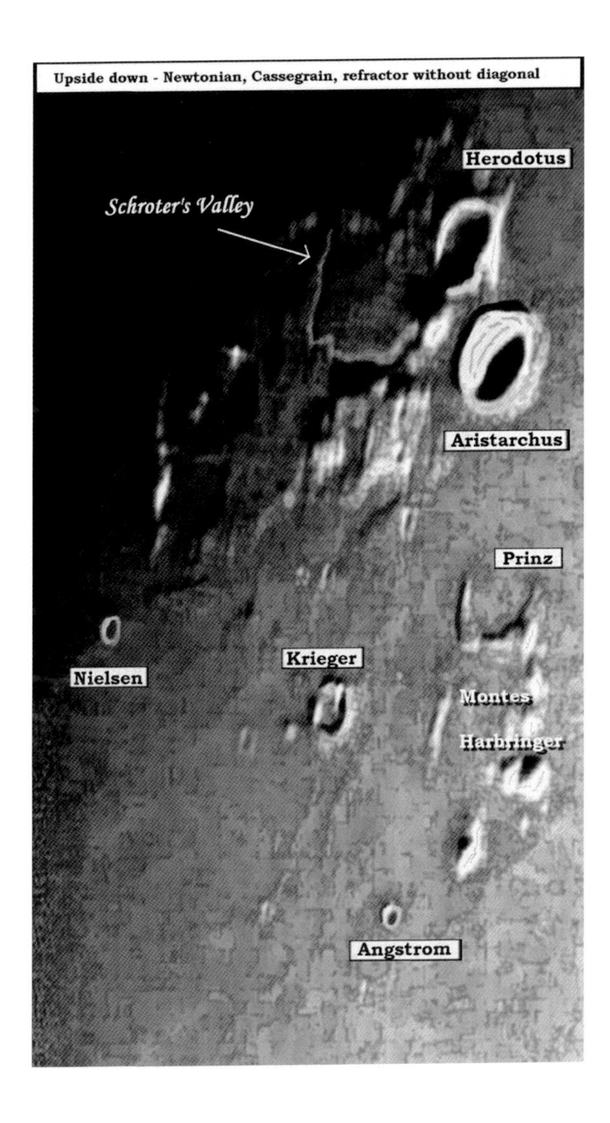
Upside down - Newtonian, Cassegrain, refractor without diagonal
Herodotus
Schroter's Valley
Aristarchus
Prinz
Nielsen
Krieger
Montes
Harbringer
Angstrom

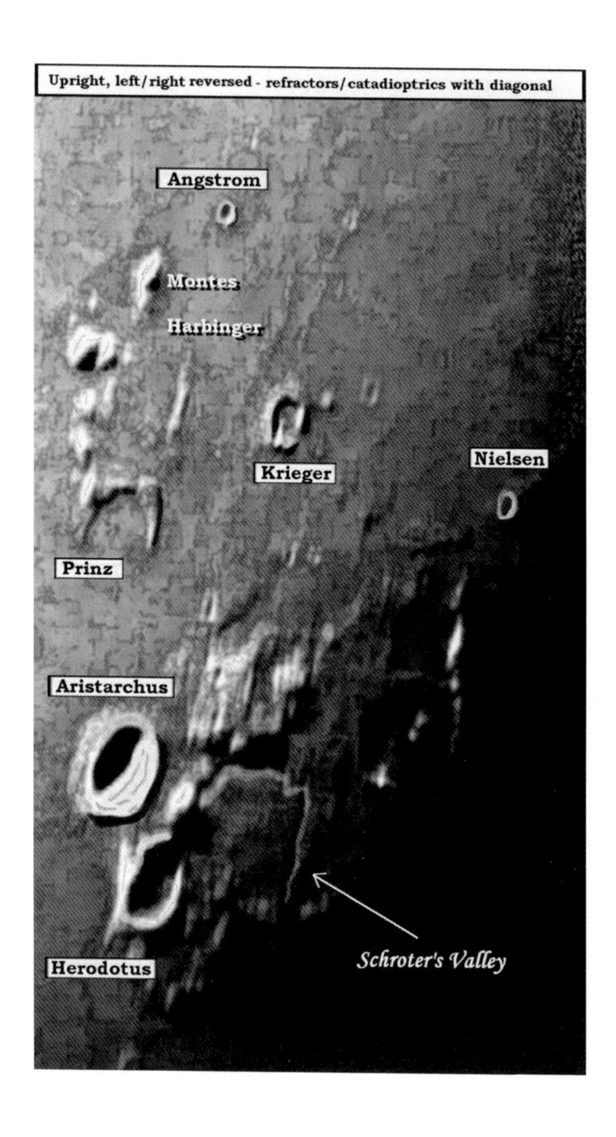
Upright, left/right reversed - refractors/catadioptrics with diagonal
Angstrom
Montes
Harbinger
Krieger
Nielsen
Prinz
Aristarchus
Herodotus
Schroter's Valley

## CHART 55

# MARIUS

Marius (25 miles) is an isolated crater in the Ocean of Storms. It is rather prominent and you will know you are viewing it when you see a large field of small hills and domes. Marius is located on the eastern edge of these hills. There is no official name for the hills, but they are usually called the Marius Hills. Under low solar illumination my 2.4 inch refractor will reveal dozens of the hills and domes. In my 6 inch refractor I would guess there are over 200. This is a fascinating area 4 days after first quarter. Explore this region at your next opportunity, you will not be disappointed.

D. Spain, *The Six-Inch Lunar Atlas: A Pocket Field Guide*,
DOI 10.1007/978-0-387-87610-8_57, 

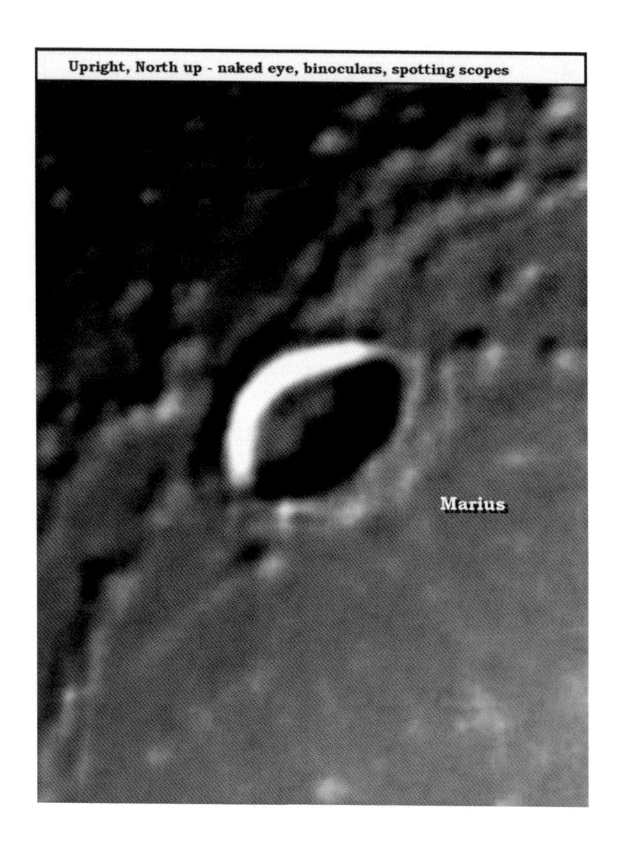
Upright, North up - naked eye, binoculars, spotting scopes
Marius

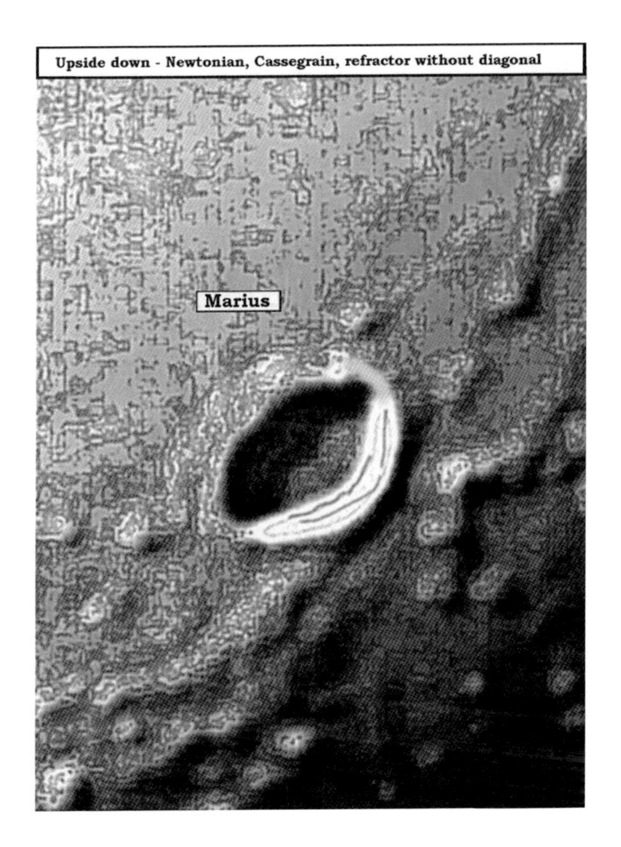
Upside down - Newtonian, Cassegrain, refractor without diagonal
Marius

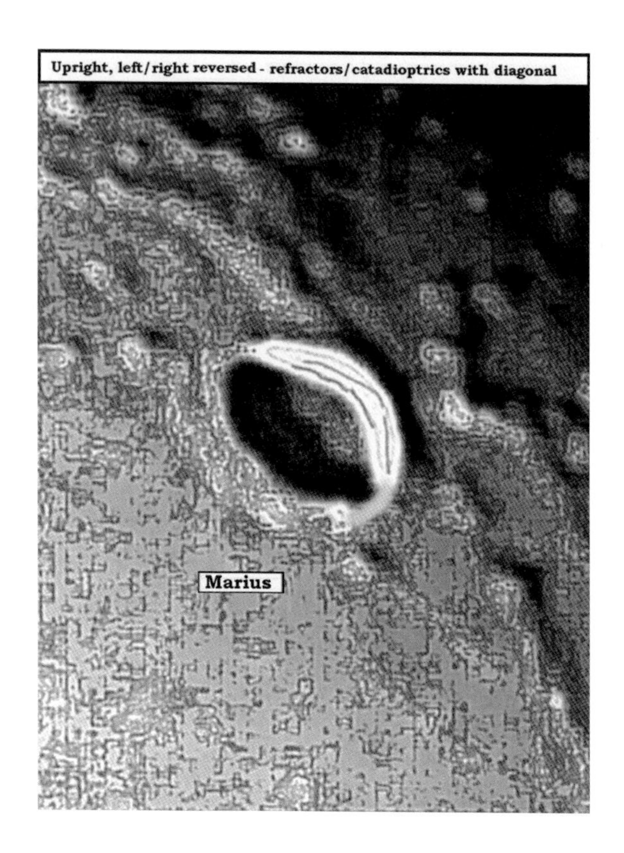
Upright, left/right reversed - refractors/catadioptrics with diagonal
Marius

## CHART 56

# HANSTEEN

Hansteen (27 miles) is a shallow crater with walls a little higher than ½ mile. The floor is rough and has some small hills. This formation sits on a slightly upraised area in the extreme southwest area of Oceanus Procellarum, the Ocean of Storms. Look for it about 4 days after first quarter. The following three formations are a little further west and best viewed a day later then Hansteen.

Sirsalis (25 miles) is a prominent crater west and north of Hansteen. On the photo it is the just above its label and is shadow filled. I estimate it at about 1 mile deep and there is a central mountain.

Damoiseau (22 miles) is fine crater just off the western shore line of the Ocean of Storms. It is about ¾ mile deep and has a little central peak.

Hermann (10 miles) is an isolated crater east and north of Damoiseau. It is easy to spot with my 2.4 inch refractor and is about ¾ mile deep.

A casual glance of this region may not look interesting, but if you explore carefully you will find partially buried formations, wrinkle ridges and isolated hills.

D. Spain, *The Six-Inch Lunar Atlas: A Pocket Field Guide*,
DOI 10.1007/978-0-387-87610-8_58, 

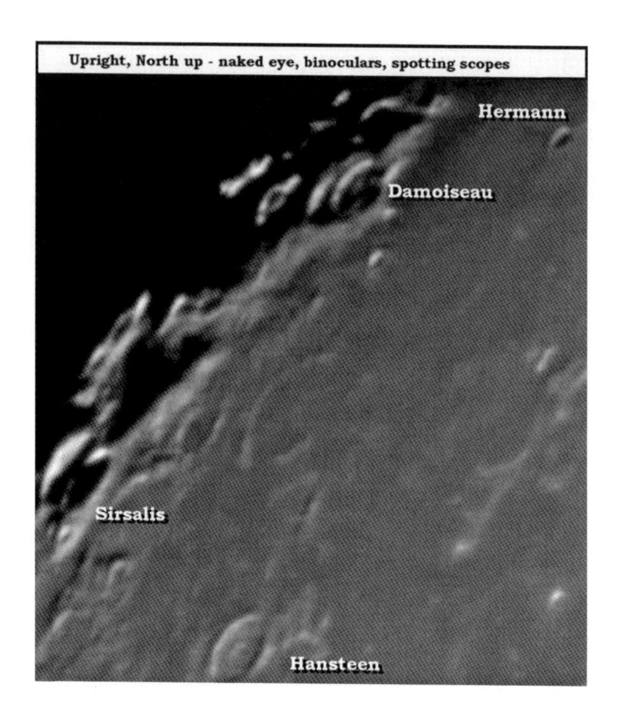
Upright, North up - naked eye, binoculars, spotting scopes
Hermann
Damoiseau
Sirsalis
Hansteen

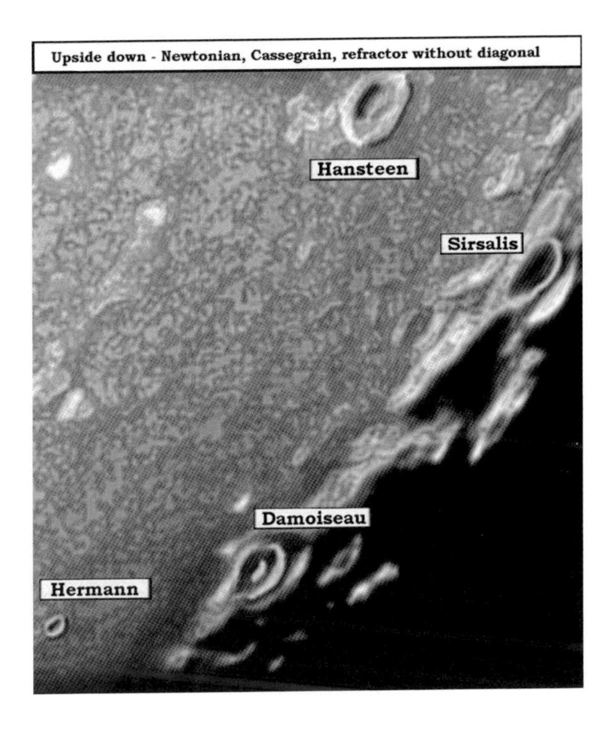
Upside down - Newtonian, Cassegrain, refractor without diagonal
Hansteen
Sirsalis
Damoiseau
Hermann

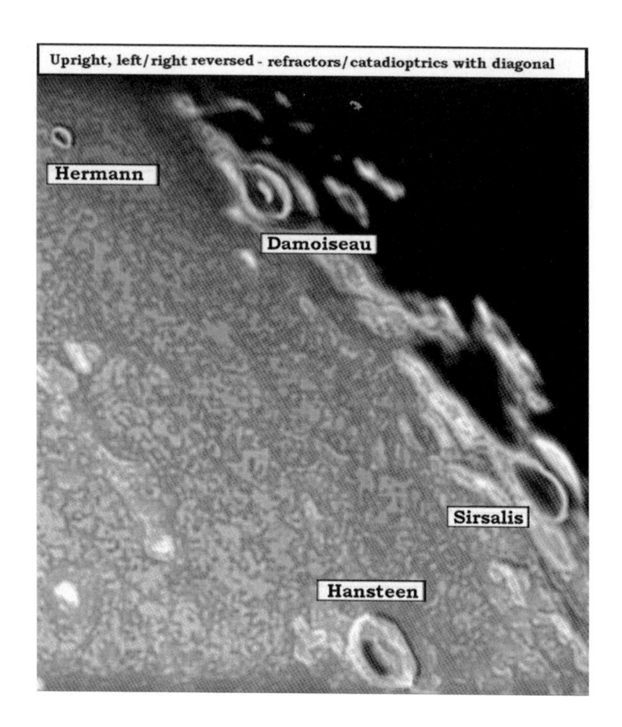
Upright, left/right reversed - refractors/catadioptrics with diagonal
Hermann
Damoiseau
Sirsalis
Hansteen

## CHART 57

# MERSENIUS

Mersenius (50 miles) is a fine formation with terraced walls over 1 mile high. There is a 9 mile crater on its southern wall that intrudes on the otherwise well formed rim. When the sun illuminates the floor 4 days after first quarter you will notice that the floor is distinctly convex. Also, notice that the floor is darker on the western side with exaggerates the convex appearance of the floor.

Liebig (22 miles) is south of Mersenius and is almost a mile deep. As the sun lights the inside of the eastern rim a 7 mile crater is revealed.

De Gasparis (18 miles) is a shallow crater with walls that are less than 1 mile high. The rim is lower in the north and under very low illumination there are several shallow rills outside the walls. Under ideal conditions I can make out a rill running the length of the floor.

Cavendish (34 miles) is a fairly deep with walls over 1 mile above the floor. There are two craters on its rim, a small 6 mile crater on the northeast and a 14 mile crater on the southwest.

D. Spain, *The Six-Inch Lunar Atlas: A Pocket Field Guide*,
DOI 10.1007/978-0-387-87610-8_59, 

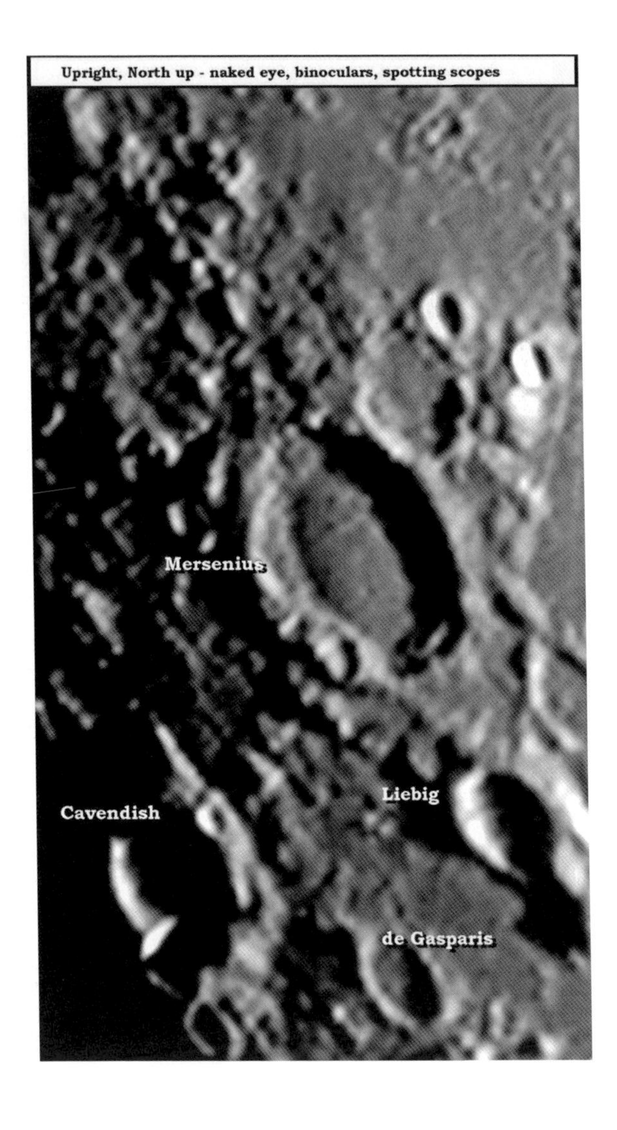
Upright, North up - naked eye, binoculars, spotting scopes
Mersenius
Liebig
Cavendish
de Gasparis

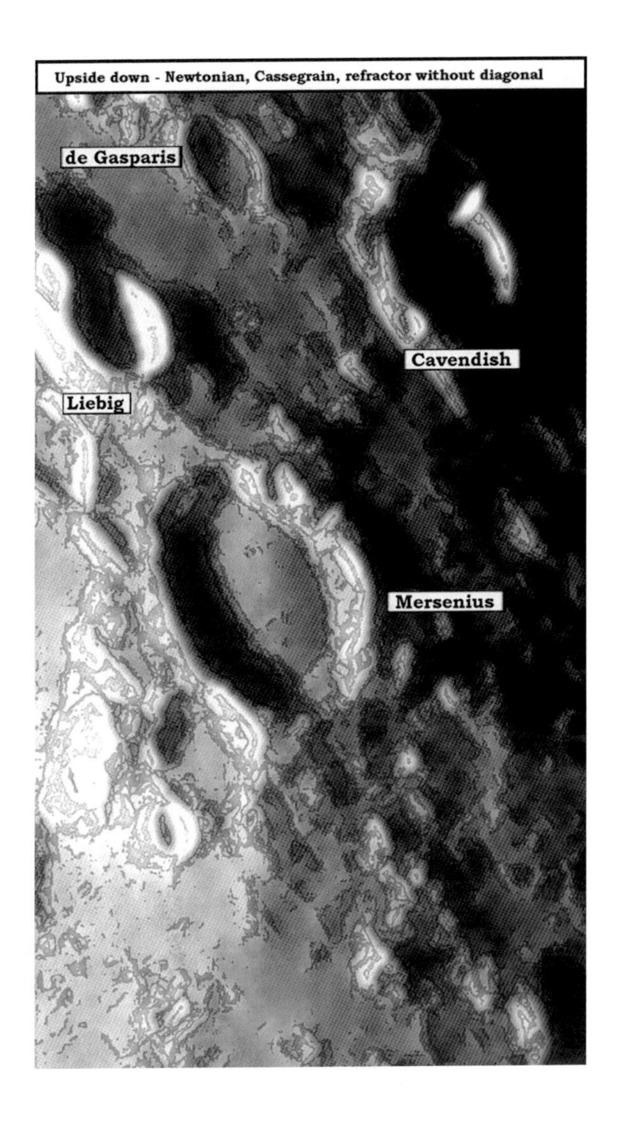
Upside down - Newtonian, Cassegrain, refractor without diagonal
de Gasparis
Cavendish
Liebig
Mersenius

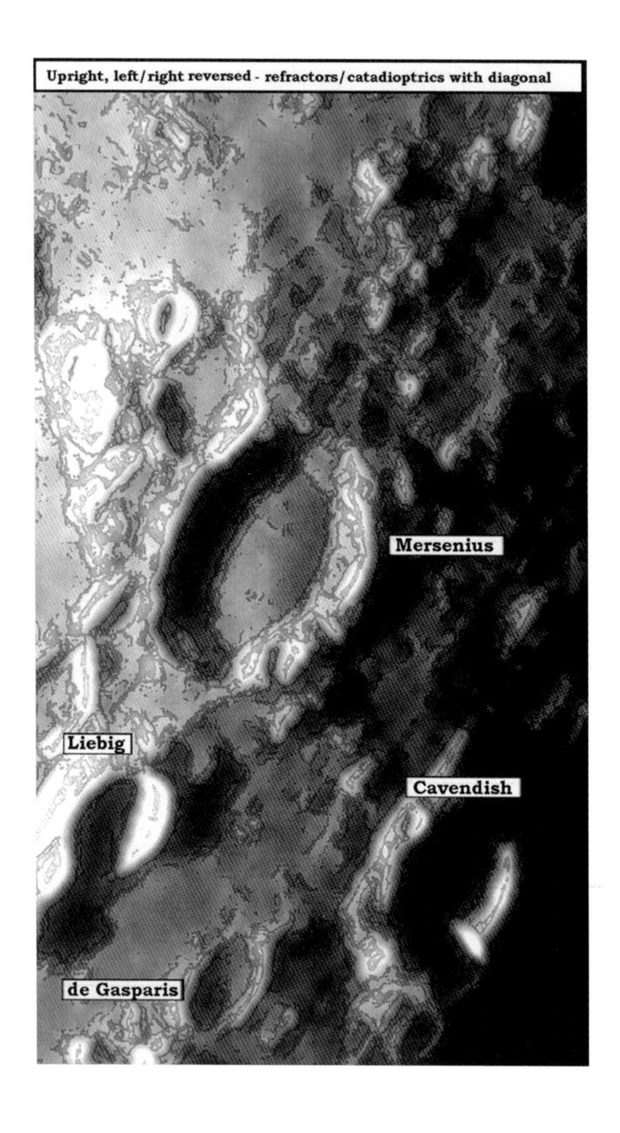
Upright, left/right reversed - refractors/catadioptrics with diagonal
Mersenius
Liebig
Cavendish
de Gasparis

## CHART 58

# RUMKER

Rumker has the appearance of a blister on the lunar surface in small scopes. In my 6 inch refractor several bumps or domes are visible. The formation is really a great complex of domes on a raised plateau. The diameter is about 40 miles. It is visible in my 2.4 inch just as the sun is rising on this formation 5 days after first quarter. If you examine Rumker carefully you will see that part of the eastern area is lower than the rest of the formation. This area is in the shape of a diamond and is easily visible on the photo. As the sun gets higher Rumker all but disappears from view.

Naumann (6 miles) is a sharp rimmed isolated crater. It is prominent and rather deep.

There is another prominent 6 mile crater between Naumann and Rumker that is unnamed. It sits at the end of a wrinkle ridge. You will see several other wrinkle ridges throughout this region.

D. Spain, *The Six-Inch Lunar Atlas: A Pocket Field Guide*,
DOI 10.1007/978-0-387-87610-8_60, 

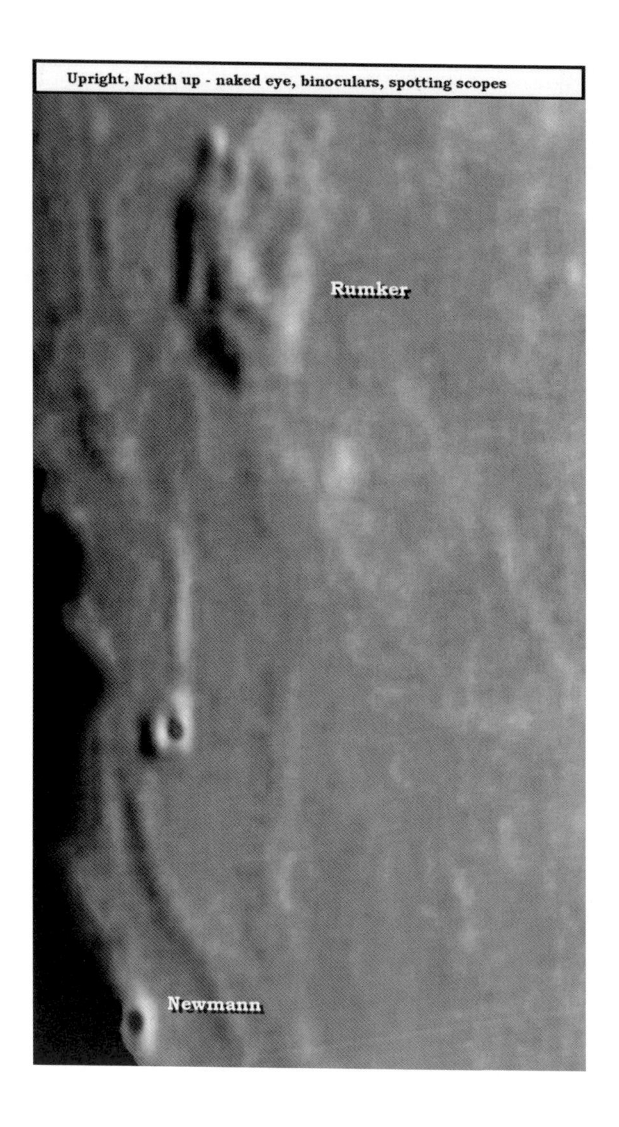
Upright, North up - naked eye, binoculars, spotting scopes
Rumker
Newmann

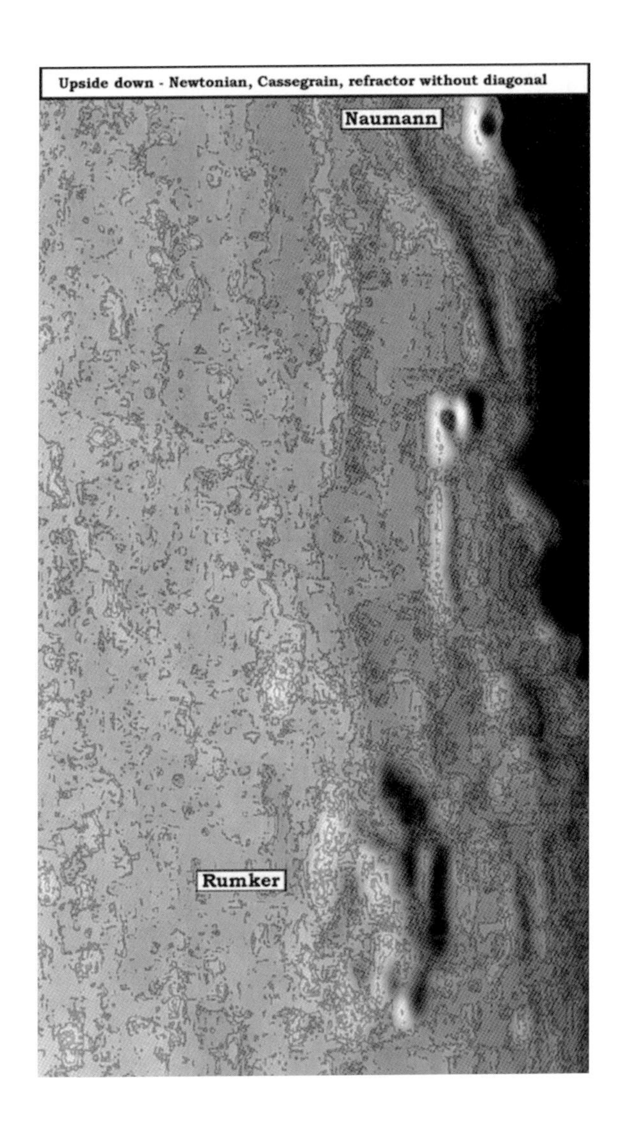
Upside down - Newtonian, Cassegrain, refractor without diagonal
Naumann
Rumker

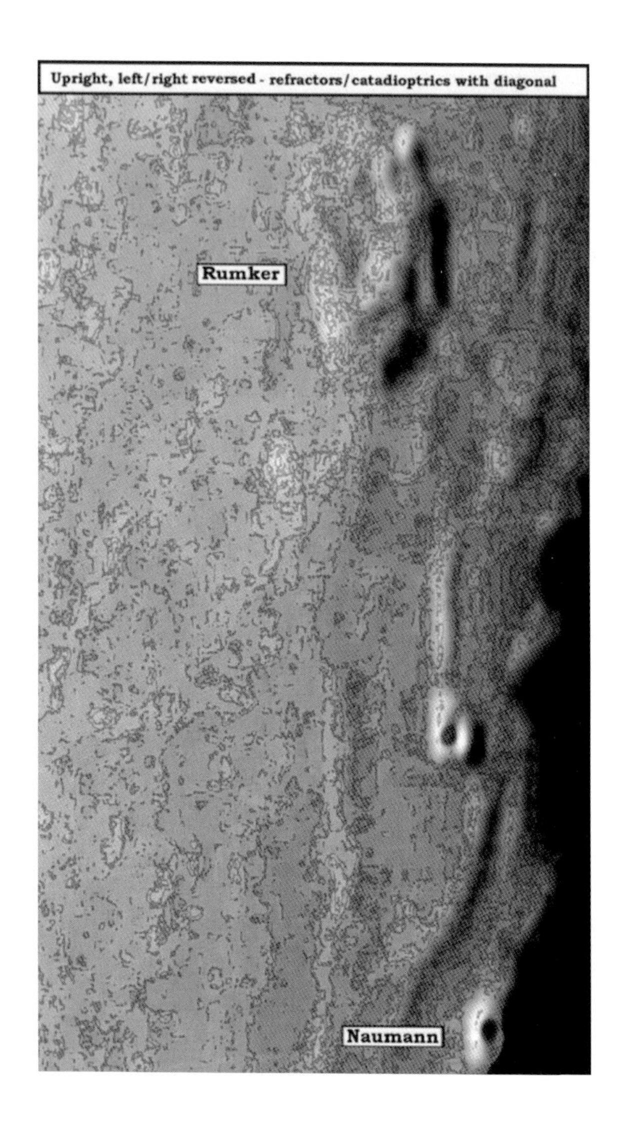
Upright, left/right reversed - refractors/catadioptrics with diagonal
Rumker
Naumann

## CHART 59

# GRIMALDI

Grimaldi (134 miles) is a great dark floored formation. The dark floor is lava that welled up after a great impact a few billion years ago. When observing Grimaldi 6 days after first quarter you will see that the rim is missing or very low along the northeastern edge. There is a 13 mile crater on the northern floor. The large flat floor has a few craterlets and raised areas. The dark floor is easily visible with the slightest optical aid.

Lohrmann (19 miles) is a prominent crater just north of Grimaldi and touching the southern rim of Hevelius. The floor appears flat and featureless even in my 6 inch refractor.

Hevelius (64 miles) is another great crater about ½ the size of Grimaldi. There is a small central peak and my 6 inch refractor shows a complex rill system crisscrossing the floor.

D. Spain, *The Six-Inch Lunar Atlas: A Pocket Field Guide*,
DOI 10.1007/978-0-387-87610-8_61, 

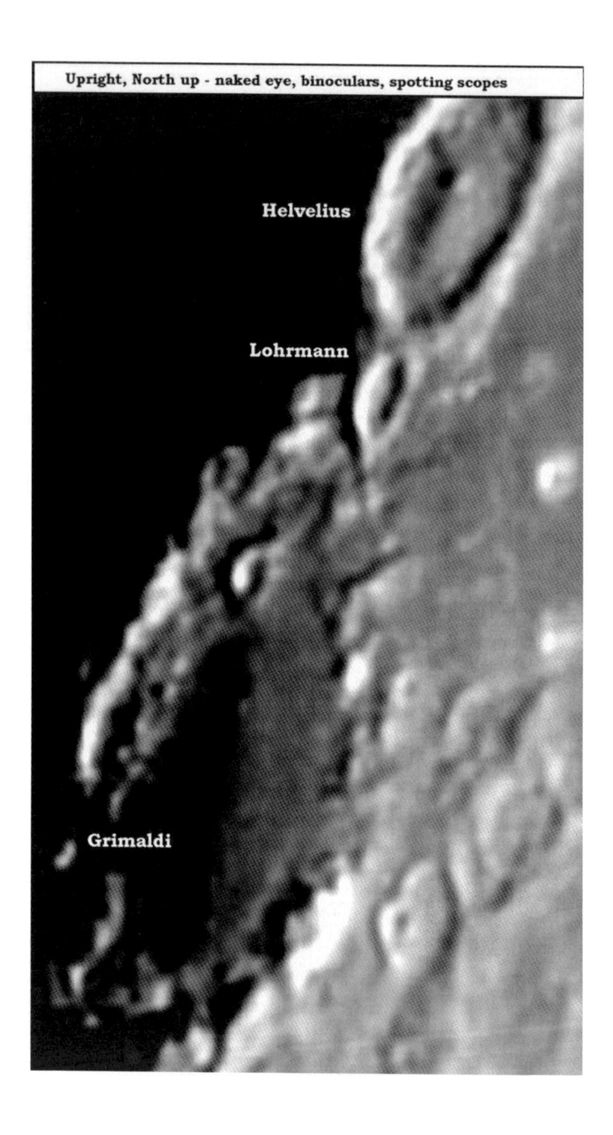
Upright, North up - naked eye, binoculars, spotting scopes
Helvelius
Lohrmann
Grimaldi

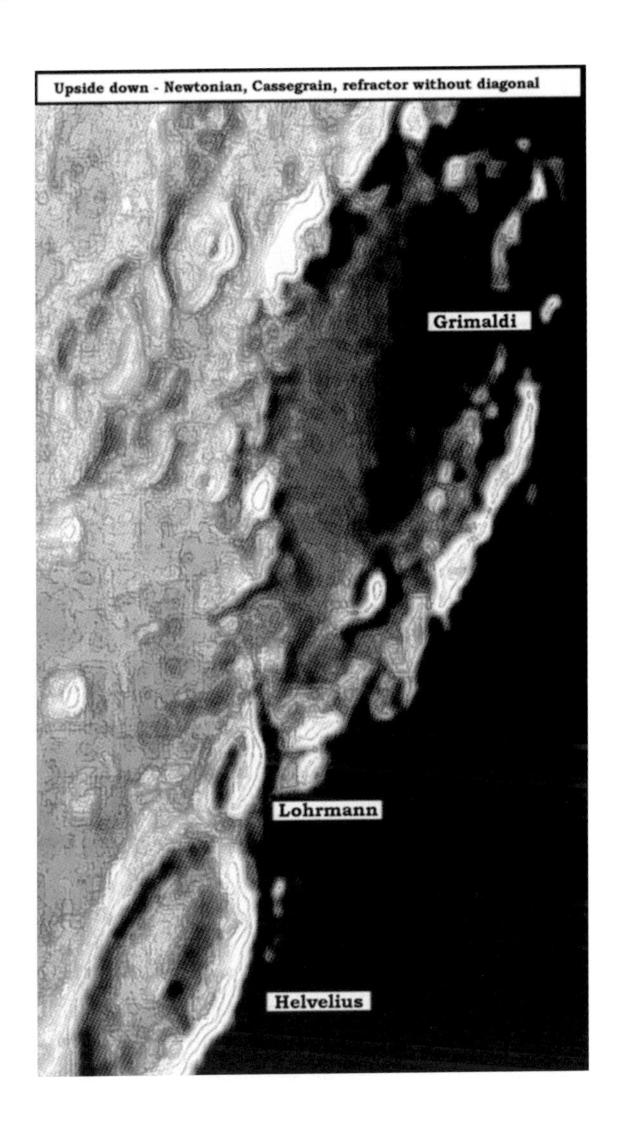
Upside down - Newtonian, Cassegrain, refractor without diagonal
Grimaldi
Lohrmann
Helvelius

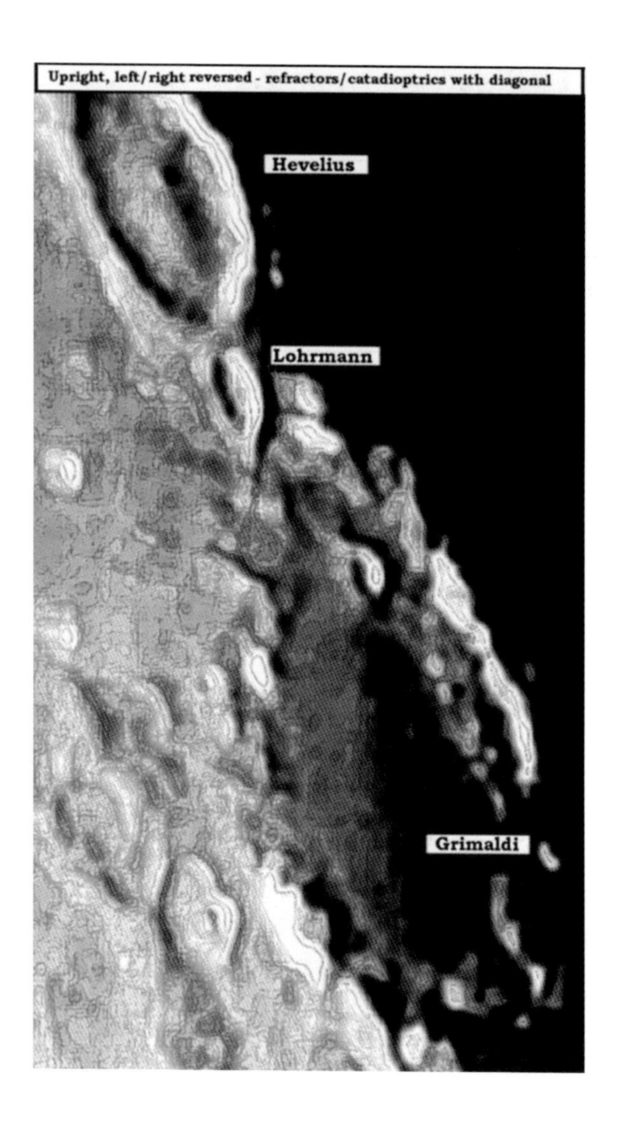
Upright, left/right reversed - refractors/catadioptrics with diagonal
Hevelius
Lohrmann
Grimaldi

## CHART 60

# STRUVE

Struve (100 miles) is another old lunar giant. Its walls are low and the vast floor is flat. It merges with Russell to the north and there is no wall separating the two. There are two 9 mile craters on the floor near the eastern rim. Under low light some low crests and a few craterlets are visible on the flooded floor.

Russell (64 miles) is Struve's companion to the north. The walls are higher in the north and there is a well formed 14 mile crater on the eastern rim. Some small craters are scattered on the large flooded floor.

Eddington (80 miles) is another flooded crater. The walls are very low, except along the western rim against Struve. Parts of the rim are missing altogether on the south and east. The floor is smooth although there are a few low crests visible under a very low sun.

Briggs (22 miles) is a fine crater out on the plain. Its walls are a little over ½ mile high and they encompass a rough floor with a central peak.

Seleucus (25 miles) is the deepest crater in this area. It is over a mile deep with a central hill. It sits off a great curving ray that passes just by its eastern rim.

This region is usually bypassed by many observers just because it is visible 1 day or less before the full Moon. However, this trio of the three giants would appreciate the occasional visit from a dedicated lunar observer. Most public observations are rarely held near a full Moon, but if somehow you happen to be at one, be sure to point out these old gentlemen.

D. Spain, *The Six-Inch Lunar Atlas: A Pocket Field Guide*,
DOI 10.1007/978-0-387-87610-8_62, 

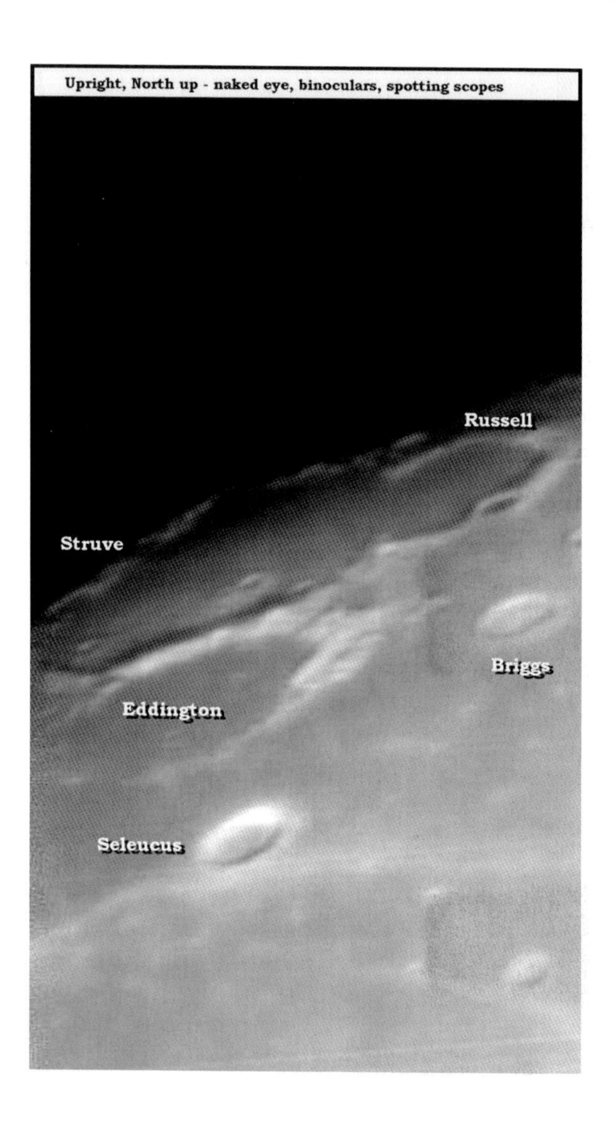
Upright, North up - naked eye, binoculars, spotting scopes
Russell
Struve
Briggs
Eddington
Seleucus

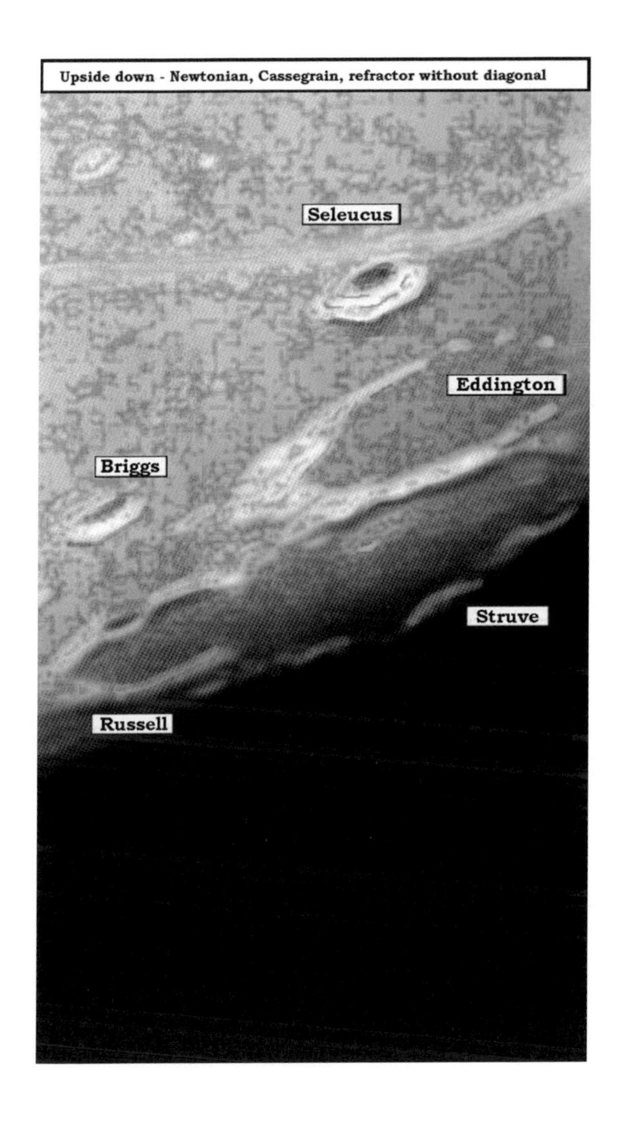
Upside down - Newtonian, Cassegrain, refractor without diagonal
Seleucus
Eddington
Briggs
Struve
Russell

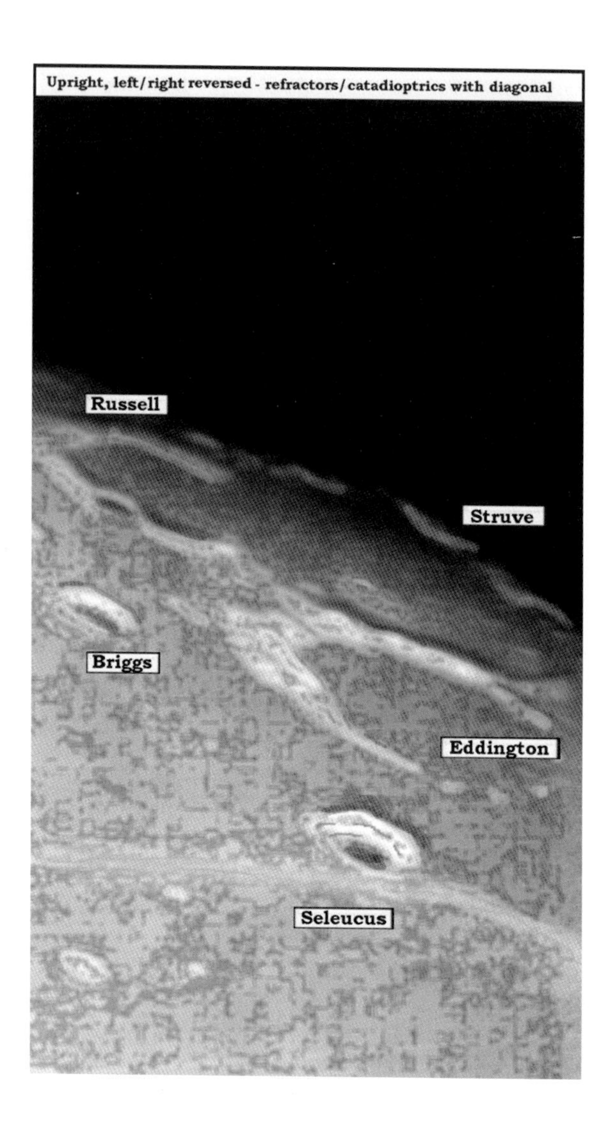
Upright, left/right reversed - refractors/catadioptrics with diagonal
Russell
Struve
Briggs
Eddington
Seleucus

# Resource

For those who want to investigate and explore the Moon further I am listing just one resource. In my opinion it is the best resource available for the lunar observer. It is the Virtual Lunar Atlas. It is completely free and is an invaluable tool.

It shows the current phase of the Moon and all important craters on the near side. By clicking on the formation you will get valuable data about the feature. I often use it to plan targets for star parties and public observations. You can go backward and forward in time by using the ephemeris and setting whatever date and time you wish.

There is a section for recording notes of your observations. There is a terminator tab that shows features of interests along the sunrise or sunset line. In the tools section there is a neat tool that lets you measure distances between points on the lunar surface. There are other features that I will leave to the reader to discover.

You may set up the program to include different catalogs of lunar photographs that are available at the download site and they are also entirely free. If you are, or decide to become a lunar photographer you can place your own photos in the program. The download site is http://ap-i.net/avl/en/start. For anyone with comments, questions, or just help with observing the Moon, feel free to contact me by email at tiphiid@yahoo.com.

# Index